T0135931

Augsburger Schriften zur Mathematik, Physik und Informatik
Band 30

Bibliografische Information der Deutschen Nationalbibliothek

Die Deutsche Nationalbibliothek verzeichnet diese Publikation in der Deutschen Nationalbibliografie; detaillierte bibliografische Daten sind im Internet über http://dnb.d-nb.de abrufbar.

ISBN 978-3-8325-4084-5
ISSN 1611-4256

Logos Verlag Berlin GmbH
Comeniushof, Gubener Str. 47,
10243 Berlin
Tel.: +49 030 42 85 10 90
Fax: +49 030 42 85 10 92
INTERNET: http://www.logos-verlag.de

Quiver $\mathcal{D}$-Modules and the Riemann-Hilbert Correspondence

Dissertation

zur Erlangung des akademischen Grades
Dr. rer. nat.

eingereicht an der
Mathematisch-Naturwissenschaftlich-Technischen Fakultät
der
Universität Augsburg

von

Stephanie Zapf

1. Gutachter: Prof. Dr. Marco Hien
2. Gutachter: Prof. Dr. Marc Nieper-Wißkirchen
Tag der mündlichen Prüfung: 14. Juli 2015

Contents

Introduction

In 1900, on the Second International Congress of Mathematicians in Paris, David Hilbert announced twenty-three problems "from the discussion of which an advancement of science may be expected" [Hil02]. These famous, so called Hilbert problems cover several main branches of mathematics like analysis, geometry, number theory and algebra, as well as foundations of mathematics. Since then Hilbert's list[1] has attracted many mathematicians hoping to solve one of the open questions and to gain reputation. In the following we will focus on the twenty-first problem which has already been solved. It deals with linear differential equations in one variable and is in Hilbert's words "an important problem, one which very likely Riemann himself may have had in mind" [Hil02]. Apparently, this problem and its generalizations are now often called Riemann-Hilbert problem or Riemann-Hilbert correspondence. It evokes the issue

> *"To show that there always exists a linear differential equation of the Fuchsian class, with given singular points and monodromic group."* [Hil02]

in which the Fuchsian equation is commonly interpreted as Fuchsian system. In 1989 a counterexample to this precise formulation by Hilbert was discovered by Andrei Andreevich Bolibruch which refuted a former proof of the assertion by Josib Plemelj in 1908 (see [AB94] and [Ple08]). Plemelj proved that a regular system of the desired type exists but his proof of the existence of a Fuchsian system failed. In fact, every Fuchsian system, i. e. a linear differential system with finitely many simple poles in $\overline{\mathbb{C}}$, is a regular one but not vice versa.

A generalization of Plemelj's statement which works in higher dimension as well stems from Pierre Deligne in 1970 (see [Del70]). Given a smooth, connected algebraic variety X over $\mathbb{C}$, Deligne proved a type of Riemann-Hilbert equivalence between the category of local systems of finite dimensional vector spaces over $\mathbb{C}$ on the analytic space X^{an} and the category of vector bundles on X equipped with a regular integrable connection ∇. Here the second category plays the role of the regular differential system in Plemelj's theorem.

In the meantime many similar but different statements yield the name Riemann-Hilbert correspondence. The general version which – in the theory of $\mathscr{D}$-modules – is known as the Riemann-Hilbert correspondence was proven independently by Masaki Kashiwara (see [Kas84]) and Zoghman Mebkhout (see [Meb84]) in 1980.[2] This fundamental result yields an equivalence between the derived categories $\mathrm{D}^b_{\mathrm{rh}}(\mathscr{D}_X)$ and $\mathrm{D}^b_{c}(\mathbb{C}_X)$. In particular this gives an equivalence

$$\mathcal{M}od_{\mathrm{rh}}(\mathscr{D}_X) \xrightarrow{\cong} \mathrm{Perv}(X)$$

between regular holonomic $\mathscr{D}_X$-modules and perverse sheaves on X. The fact that this theorem, as well as Deligne's correspondence, links algebra/analysis with topology, makes it so striking.

[1]For some introductions to Hilbert's twenty-three problems and their solutions see for instance [Gra00], [Thi05] and [Yan02]. But this list makes no claim to be representative at all.

[2]Actually, a more general version of the Riemann-Hilbert correspondence was presented by Andrea d'Agnolo and Masaki Kashiwara in November 2013. They give a correspondence for all holonomic $\mathscr{D}$-modules and not only for regular ones.

Now, let us come back to dimension one locally at 0. In this situation the local system in Deligne's theorem is equivalent to the data of a n-dimensional $\mathbb{C}$-vector space and an endomorphism of $\mathbb{C}^n$. This gives rise to an equivalent formulation of his statement. In more generality one obtains an equivalence of the category $\mathcal{M}od_{\mathrm{rh}}(\mathscr{D})$ of regular holonomic $\mathscr{D}$-modules with the category $\mathcal{C}_1$ of finite quiver representations $E \underset{v}{\overset{u}{\rightleftarrows}} F$ over $\mathbb{C}$ fulfilling that $u \circ v + \mathrm{Id}$ is invertible. This statement can be verified by showing that the category of perverse sheaves is equivalent to the category $\mathcal{C}_1$ of such quiver representations in this situation. Equivalently, the statement might be proven directly without the use of perverse sheaves by giving a pair of quasi-inverse functors and in particular a functor from $\mathcal{C}_1$ into $\mathcal{M}od_{\mathrm{rh}}(\mathscr{D})$.

Obviously, the category $\mathcal{C}_1$ is easy to handle. However, it arises the question of higher dimensions. A paper of André Galligo, Michel Granger and Philippe Maisonobe goes exactly in this direction (see [GGM85a]). They prove that in the case of a normal crossing divisor in dimension n, the category of perverse sheaves with respect to the induced normal crossing stratification is equivalent to the category $\mathcal{C}_n$ ($\mathcal{C}_n$ is the generalization of $\mathcal{C}_1$). This means, using the Riemann-Hilbert correspondence again, that in this normal crossing situation (see [GGM85b])

$$\mathcal{M}od^{S}_{\mathrm{rh}}(\mathscr{D}) \underset{\cong}{\xrightarrow{\mathcal{A}}} \mathcal{C}_n\,,$$

i. e. the category of regular holonomic $\mathscr{D}$-modules whose singular locus is a normal crossing is equivalent to $\mathcal{C}_n$. However, in contrast to dimension one, in higher dimension it is not that easy to assign a $\mathscr{D}$-module to a given quiver representation with respect to this equivalence concretely.

A contribution to the question of how to assign a $\mathscr{D}$-module to a quiver representation comes from Sergei Khoroshkin and Alexander Varchenko (see [KV06]). To a given hyperplane arrangement in $\mathbb{C}^n$, they associate a quiver. And to each finite representation over $\mathbb{C}$ of such a quiver (without any extra conditions on the representation), they associate a $\mathscr{D}$-module in a rather intuitive way. This yields a functor E from the category of representations over these quivers into the category of holonomic $\mathscr{D}$-modules. Using this definition in dimension one, one sees that this gives, up to an isomorphism, exactly the functor from $\mathcal{C}_1$ to $\mathcal{M}od_{\mathrm{rh}}(\mathscr{D})$ from above. In particular one sees that every regular holonomic $\mathscr{D}$-module in dimension one locally at 0 is isomorphic to a quiver $\mathscr{D}$-module. This makes their construction very promising for higher dimensions.

Now, the main idea of this work is to use this construction of quiver $\mathscr{D}$-modules by Khoroshkin and Varchenko in the case of a normal crossing hyperplane arrangement and to combine it with the theorem of Galligo, Granger and Maisonobe. We show that in this situation the functor E maps from the category $\mathcal{Q}ui_n$ of representations of the "hypercube quiver" into the category $\mathcal{M}od^{S}_{\mathrm{rh}}(\mathscr{D})$. After restricting the category $\mathcal{Q}ui_n$ (which yields the category $\mathcal{Q}ui_n^{\Sigma_1}$) we obtain that

$$E\colon \mathcal{Q}ui_n^{\Sigma_1} \to \mathcal{M}od^{S}_{\mathrm{rh}}(\mathscr{D})$$

is an equivalence of categories. In particular, this yields that every $\mathscr{D}$-module in $\mathcal{M}od^{S}_{\mathrm{rh}}(\mathscr{D})$ is in fact isomorphic to a quiver $\mathscr{D}$-module. More precisely, we will show in our Main Theorem 4.5 that the following diagram commutes up to a natural isomorphism

$$\begin{array}{ccc} \mathcal{M}od^{S}_{\mathrm{rh}}(\mathscr{D}) & \xrightarrow{\mathscr{A}} & \mathcal{C}_n \\ {\scriptstyle E}\uparrow & & \uparrow{\scriptstyle Q} \\ \mathcal{Q}ui_n^{\Sigma_1} & \xrightarrow{D} & \mathcal{Q}ui_n^{\Sigma_1} \end{array}$$

where all the functors involved are equivalences of categories.

The present work is organized as follows:

In Part I of this work we prove that the categories $\mathcal{C}_n$ and $\mathcal{Q}ui_n^{\Sigma_1}$ of (hypercube) quiver representations are equivalent. First, we are going to state the main definitions and properties of these categories. Afterwards, we prove the equivalence by stating two possible pairs of quasi-inverse functors.

Part II is devoted to the quiver $\mathscr{D}$-modules:
In Chapter 2 we are going to define $\mathscr{D}$-modules associated to a hypercube quiver representation. We show that this yields a functor E from the category $\mathcal{Q}ui_n$ into the category $\mathcal{M}od_{\mathrm{rh}}^S(\mathscr{D})$. Furthermore, we clarify the connection with [KV06].
Chapter 3 deals with the 1-dimensional case. First of all, we recapitulate the known statements on the Riemann-Hilbert correspondence in dimension one locally at 0. Then, we reprove the statement that every regular holonomic $\mathscr{D}$-module is isomorphic to a quiver $\mathscr{D}$-module in dimension one. This proof already is a special case of our main theorem and provides an idea of how the general version works. Finally, we compute a complete list of quiver $\mathscr{D}$-modules which yields the list of Boutet de Monvel (see [Bou83]).
In the final Chapter 4 we prove our main theorem, i. e. that the above square commutes up to a natural isomorphism.

Acknowledgements

First and foremost, I would like to thank my supervisor Marco Hien for supporting me throughout my research work. I am grateful for all the encouragement and vast hours of helpful discussions you spent with me and which motivated me over the years and made it possible to finish this thesis. Furthermore, it's a pleasure to thank Michel Granger for giving me the opportunity to do a residence for research in Angers with plenty of discussions and suggestions. Finally, I want to thank my colleagues, in particular Giovanni Morando and Hedwig Heizinger, for various helpful discussions and my family and friends for their constant support over the last years.

Part I.

Quiver representations

1. Finite representations of a kind of hypercube quiver

In this chapter we will consider finite representations over $\mathbb{C}$ of the following quiver:
Let $n \in \mathbb{N}^+$ and let $\mathcal{P}(\{1,\dots,n\})$ denote the power set of $\{1,\dots,n\}$. The quiver consists of 2^n vertices which we denote by $I \in \mathcal{P}(\{1,\dots,n\})$, and $n2^n$ oriented edges

$$I \rightleftarrows I \cup \{i\}$$

for $i \in \{1,\dots,n\} \setminus I$. This gives us a kind of hypercube quiver by imaging that the vertices of the quiver lie on the vertices of a n-dimensional hypercube, and we have two edges exactly for every edge of the hypercube.

In Section 1.1 we will define three categories of such representations and we give their basic properties. In Section 1.2 we show that two of these categories are isomorphic. These categories will be the fundamental building blocks of the quiver $\mathscr{D}$-modules in Part II.

1.1. Definitions and basic properties

In the following we are going to define three standard categories of hypercube quiver representations, denoted $\mathcal{Q}ui_n$, $\mathcal{C}_n$ and $\mathcal{Q}ui_n^{\Sigma_1}$. Let us start with the definition of $\mathcal{Q}ui_n$. This is basically just the category of finite representations over $\mathbb{C}$ of the above hypercube quiver.

Definition 1.1.
The category $\mathcal{Q}ui_n$ for $n \in \mathbb{N}^+$ is defined as follows:

- *The objects consist of 2^n finitely generated $\mathbb{C}$-vector spaces denoted V_I where $I \in \mathcal{P}(\{1,\dots,n\})$, equipped with $n2^n$ $\mathbb{C}$-linear mappings $u_{I,i}$ and $y_{I,i}$ for $i \in \{1,\dots,n\} \setminus I$,*

$$V_I \underset{y_{I,i}}{\overset{u_{I,i}}{\rightleftarrows}} V_{I\cup\{i\}}\,,$$

and they satisfy the following commutativity conditions for $i,j \in \{1,\dots,n\} \setminus I$:

$$u_{I\cup\{i\},j} \circ u_{I,i} = u_{I\cup\{j\},i} \circ u_{I,j} \qquad y_{I,i} \circ y_{I\cup\{i\},j} = y_{I,j} \circ y_{I\cup\{j\},i}$$
$$y_{I\cup\{i\},j} \circ u_{I\cup\{j\},i} = u_{I,i} \circ y_{I,j} \qquad u_{I,j} \circ y_{I,i} = y_{I\cup\{j\},i} \circ u_{I\cup\{i\},j}$$

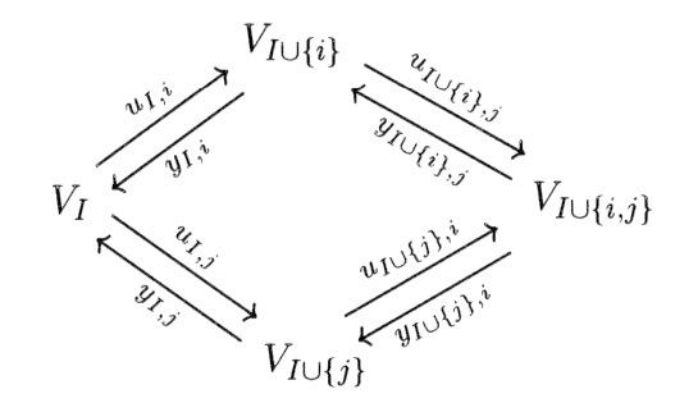

- *A morphism between two objects* $(V_I, u_{I,i}, y_{I,i})$ *and* $(V'_I, u'_{I,i}, y'_{I,i})$ *is given by* 2^n *linear maps* $h_I\colon V_I \to V'_I$ *such that* $u'_{I,i} \circ h_I = h_{I\cup\{i\}} \circ u_{I,i}$ *and* $h_I \circ y_{I,i} = y'_{I,i} \circ h_{I\cup\{i\}}$ *for* $i \in \{1,\dots,n\}\backslash I$.

$$\begin{array}{ccc} V_I & \underset{y_{I,i}}{\overset{u_{I,i}}{\rightleftarrows}} & V_{I\cup\{i\}} \\ {\scriptstyle h_I}\downarrow & & \downarrow{\scriptstyle h_{I\cup\{i\}}} \\ V'_I & \underset{y'_{I,i}}{\overset{u'_{I,i}}{\rightleftarrows}} & V'_{I\cup\{i\}} \end{array}$$

- *The identity morphism on* $(V_I, u_{I,i}, y_{I,i})$ *is given by* $(h_I) = (\mathrm{Id}_{V_I})$.
- *The composition of the morphism* (h_I) *from* $(V_I, u_{I,i}, y_{I,i})$ *to* $(V'_I, u'_{I,i}, y'_{I,i})$ *and the morphism* (h'_I) *from* $(V'_I, u'_{I,i}, y'_{I,i})$ *to* $(V''_I, u''_{I,i}, y''_{I,i})$ *is given by* $(h'_I) \circ (h_I) = (h'_I \circ h_I)$.

We want to note that the last two commutativity conditions for objects in Definition 1.1 are obviously equivalent by interchanging i and j. So, hereafter we will only treat the necessary three conditions in our proofs.

Furthermore, we want to note that using a notation by Deligne the commutativity conditions might be all summed up to $\delta_i \circ \delta_j = \delta_j \circ \delta_i$ where $\delta_i\colon V_I \to V_{I\Delta\{i\}}$ and $I\Delta\{i\}$ is the symmetric difference of I with $\{i\}$. This notation allows to shorten many conditions but it is not useful for us as we need to distinguish between $i \in I$ and $i \notin I$ later.

Now, let us define the categories $\mathcal{C}_n$ and $\mathcal{Q}ui_n^{\Sigma_1}$. They are subcategories of $\mathcal{Q}ui_n$ fulfilling an additional constraint on their objects. The category $\mathcal{C}_n$ is used in [GGM85a] and [GGM85b] for example. For $n = 1$ the category $\mathcal{Q}ui_n^{\Sigma_1}$ is used in [Bjö93, Section V.2], denoted $\mathscr{A}$ there. We use its natural generalisation to arbitrary $n \in \mathbb{N}^+$.

Definition 1.2.
The category $\mathcal{C}_n$ *is the full subcategory of* $\mathcal{Q}ui_n$ *such that every object* $(V_I, u_{I,i}, y_{I,i})$ *additionally fulfils that* $u_{I,i} \circ y_{I,i} + \mathrm{Id}_{V_{I\cup\{i\}}}$ *and* $y_{I,i} \circ u_{I,i} + \mathrm{Id}_{V_I}$ *are invertible.*

Definition 1.3.
The category $\mathcal{Q}ui_n^{\Sigma_1}$ *is the full subcategory of* $\mathcal{Q}ui_n$ *such that every object* $(V_I, u_{I,i}, y_{I,i})$ *additionally fulfils that the eigenvalues of* $u_{I,i} \circ y_{I,i}$ *and* $y_{I,i} \circ u_{I,i}$ *lie in* $\Sigma_1 := \Sigma + 1$ *where*

$$\Sigma := \left\{ \alpha \in \mathbb{C} \;\middle|\; -1 \leq \mathrm{Re}(\alpha) \leq 0,\ \mathrm{Im}(\alpha) = \begin{cases} \geq 0 & \textit{if } \mathrm{Re}(\alpha) = -1, \\ < 0 & \textit{if } \mathrm{Re}(\alpha) = 0, \\ \textit{arbitrary} & \textit{otherwise.} \end{cases} \right\}$$

In the remaining part of this section we are going to state and prove some well-known properties concerning our three categories. The following statement helps us to shorten the definitions of the categories $\mathcal{C}_n$ and $\mathcal{Q}ui_n^{\Sigma_1}$.

Proposition 1.4.
Let E, F *denote two finite dimensional* $\mathbb{C}$*-vector spaces and let* $c\colon E \to F$, $v\colon F \to E$ *denote two* $\mathbb{C}$*-linear mappings. We fix a complex number* $\lambda \in \mathbb{C}$. *Then* $v \circ c + \lambda \mathrm{Id}_E$ *and* $c \circ v + \lambda \mathrm{Id}_F$ *fulfil:*

$$\mathrm{Spec}(v \circ c + \lambda \mathrm{Id}_E) \setminus \{\lambda\} = \mathrm{Spec}(c \circ v + \lambda \mathrm{Id}_F) \setminus \{\lambda\}$$

Proof:
As $\operatorname{Spec}(v \circ c + \lambda \operatorname{Id}_E) = \{\operatorname{Spec}(v \circ c) + \lambda\}$ and $\operatorname{Spec}(c \circ v + \lambda \operatorname{Id}_F) = \{\operatorname{Spec}(c \circ v) + \lambda\}$, the statement is equivalent to

$$\operatorname{Spec}(v \circ c) \setminus \{0\} = \operatorname{Spec}(c \circ v) \setminus \{0\} .$$

We prove the inclusions "$\subseteq$" and "$\supseteq$":

"$\subseteq$" Let $\mu \neq 0$ denote an eigenvalue of the map $v \circ c$ and let $e \neq 0$ denote an eigenvector corresponding to the eigenvalue μ, i. e. $(v \circ c)(e) = \mu e$. This equation shows that $c(e) \neq 0$. Moreover, we have

$$(c \circ v)(c(e)) = c(\mu e) = \mu c(e) .$$

Hence, μ is an eigenvalue of $c \circ v$ corresponding to the eigenvector $c(e)$.

"$\supseteq$" This proof is analogously to the proof of the inclusion "$\subseteq$". □

Now, we can show that invertibility of $u_{I,i} \circ y_{I,i} + \operatorname{Id}$ is equivalent to invertibility of $y_{I,i} \circ u_{I,i} + \operatorname{Id}$ which simplifies Definition 1.2:

Corollary 1.5.
Let E, F denote two finite dimensional $\mathbb{C}$-vector spaces and let $c\colon E \to F$, $v\colon F \to E$ denote two $\mathbb{C}$-linear mappings. Then the following statements are equivalent:

(i) $v \circ c + \operatorname{Id}_E$ *is invertible.*

(ii) $c \circ v + \operatorname{Id}_F$ *is invertible.*

Proof:
Using Proposition 1.4 for $\lambda = 1$, we see that 0 is an eigenvalue of $v \circ c + \operatorname{Id}_E$ if and only if 0 is an eigenvalue of $c \circ v + \operatorname{Id}_F$. This immediately yields that $v \circ c + \operatorname{Id}_E$ is invertible if and only if $c \circ v + \operatorname{Id}_F$ is invertible. □

Furthermore, we can use Proposition 1.4 to simplify Definition 1.3 by showing that the eigenvalues of $u_{I,i} \circ y_{I,i}$ lie in Σ_1 if and only if the eigenvalues of $y_{I,i} \circ u_{I,i}$ lie in Σ_1:

Corollary 1.6.
Let E, F denote two finite dimensional $\mathbb{C}$-vector spaces and let $c\colon E \to F$, $v\colon F \to E$ denote two $\mathbb{C}$-linear mappings. Then the following statements are equivalent:

(i) The eigenvalues of $v \circ c\colon E \to E$ *lie in* Σ_1.

(ii) The eigenvalues of $c \circ v\colon F \to F$ *lie in* Σ_1.

Proof:
Using Proposition 1.4 for $\lambda = 0$, we see that every non-zero eigenvalue of $v \circ c$ is an eigenvalue of $c \circ v$ and vice versa. Using the fact that $0 \in \Sigma_1$, we obtain the claimed equivalence. □

Now, we would like to clarify how isomorphisms in the category $\mathcal{Q}ui_n$, and therefore in the categories $\mathcal{C}_n$ and $\mathcal{Q}ui_n^{\Sigma_1}$, look like. This will be done in Proposition 1.7(i). It shows that a morphism (h_I) in $\mathcal{Q}ui_n$ is an isomorphism if and only if every h_I is an isomorphism, just as one would intuitively say. Apart from that, we show in Corollary 1.8 that $\mathcal{C}_n$ and $\mathcal{Q}ui_n^{\Sigma_1}$ are saturated subcategories of $\mathcal{Q}ui_n$.

Proposition 1.7.

(i) In Qui_n a morphism (h_I) from $(V_I, u_{I,i}, y_{I,i})$ to $(V_I', u_{I,i}', y_{I,i}')$ is an isomorphism if and only if every h_I is an isomorphism.

(ii) Let (h_I) denote an isomorphism in Qui_n from $(V_I, u_{I,i}, y_{I,i})$ to $(V_I', u_{I,i}', y_{I,i}')$. Then $\mathrm{Spec}(u_{I,i} \circ y_{I,i}) = \mathrm{Spec}(u_{I,i}' \circ y_{I,i}')$ and $\mathrm{Spec}(y_{I,i} \circ u_{I,i}) = \mathrm{Spec}(y_{I,i}' \circ u_{I,i}')$.

Proof:

(i) (h_I) is an isomorphism if and only if there exists a morphism $(\tilde{h}_I)$ from $(V_I', u_{I,i}', y_{I,i}')$ to $(V_I, u_{I,i}, y_{I,i})$ such that $(\tilde{h}_I)\circ(h_I) = (\mathrm{Id}_{V_I})$ and $(h_I)\circ(\tilde{h}_I) = (\mathrm{Id}_{V_I'})$. As $(\tilde{h}_I)\circ(h_I) = (\tilde{h}_I \circ h_I)$ and $(h_I) \circ (\tilde{h}_I) = (h_I \circ \tilde{h}_I)$, the statement follows immediately.

(ii) Using the fact that

$$\begin{array}{ccc} V_I & \underset{y_{I,i}}{\overset{u_{I,i}}{\rightleftarrows}} & V_{I\cup\{i\}} \\ {\scriptstyle h_I}\downarrow & & \downarrow{\scriptstyle h_{I\cup\{i\}}} \\ V_I' & \underset{y_{I,i}'}{\overset{u_{I,i}'}{\rightleftarrows}} & V_{I\cup\{i\}}' \end{array}$$

commutes and that every h_I is an isomorphism by part (i), we see that

$$u_{I,i}' = h_{I\cup\{i\}} \circ u_{I,i} \circ h_I^{-1} \quad \text{and} \quad y_{I,i}' = h_I \circ y_{I,i} \circ h_{I\cup\{i\}}^{-1} .$$

This yields

$$u_{I,i}' \circ y_{I,i}' = h_{I\cup\{i\}} \circ u_{I,i} \circ y_{I,i} \circ h_{I\cup\{i\}}^{-1} \quad \text{and} \quad y_{I,i}' \circ u_{I,i}' = h_I \circ y_{I,i} \circ u_{I,i} \circ h_I^{-1} .$$

Now, the statement can be deduced from linear algebra. □

Proposition 1.7(ii) gives us directly the saturation of the categories $\mathcal{C}_n$ and $Qui_n^{\Sigma_1}$:

Corollary 1.8.
$\mathcal{C}_n$ and $Qui_n^{\Sigma_1}$ are saturated subcategories of the category Qui_n.
I. e. if (h_I) is an isomorphism in Qui_n from $(V_I, u_{I,i}, y_{I,i})$ to $(V_I', u_{I,i}', y_{I,i}')$ and one of these objects lies in $\mathcal{C}_n$ (respectively $Qui_n^{\Sigma_1}$), then the other one is an object in $\mathcal{C}_n$ (respectively $Qui_n^{\Sigma_1}$), too.

The last topic in this section is a dualizing functor acting on our quiver categories. This functor will be used in our Main Theorem 4.5. It is defined as follows:

Definition 1.9.
The contravariant functor D: $Qui_n \to Qui_n$ is defined as follows:
Let $(V_I, u_{I,i}, y_{I,i})$ denote an object in Qui_n. We define D on objects by

$$D\left(V_I \underset{y_{I,i}}{\overset{u_{I,i}}{\rightleftarrows}} V_{I\cup\{i\}} \right) := \quad V_I^* \underset{u_{I,i}^*}{\overset{y_{I,i}^*}{\rightleftarrows}} V_{I\cup\{i\}}^*$$

where V_I^* is the dual vector space of V_I and $u_{I,i}^*$, $y_{I,i}^*$ are the dual/transpose maps of $u_{I,i}$, $y_{I,i}$.

Let (h_I) denote a morphism in $\mathcal{Q}ui_n$ from $(V_I, u_{I,i}, y_{I,i})$ to $(\tilde{V}_I, \tilde{u}_{I,i}, \tilde{y}_{I,i})$. Then we set

$$D((h_I)) := (h_I^*)$$

where h_I^* is the dual map of h_I.
Clearly, this also establishes a functor from $\mathcal{Q}ui_n^{\Sigma_1}$ to $\mathcal{Q}ui_n^{\Sigma_1}$, and from $\mathcal{C}_n$ to $\mathcal{C}_n$.

We can immediately deduce the following corollary from Definition 1.9:

Corollary 1.10.
The functor D establishes an equivalence of categories where D is its own quasi-inverse.

Proof:
The statement is well-known from linear algebra as $V_I^{**} \cong V_I$. □

1.2. An equivalence of categories

The goal of this section is to prove the following Theorem 1.11 which states the isomorphism of the categories $\mathcal{C}_n$ and $\mathcal{Q}ui_n^{\Sigma_1}$. This theorem is a principal component of the present work. It helps to establish the commutativity of the diagram in Theorem 4.5 which enables us to prove the equivalence of the categories $\mathcal{Q}ui_n^{\Sigma_1}$ and $\mathcal{M}od_{\mathrm{rh}}^S(\mathscr{D})$ – a subcategory of the category of regular singular holonomic $\mathscr{D}$-modules (see Definition 2.3) – via quiver $\mathscr{D}$-modules (see Section 2.1).

In the following we give two pairs of functors which can be used independently to show the isomorphism of $\mathcal{C}_n$ and $\mathcal{Q}ui_n^{\Sigma_1}$. The first pair Q, $\mathcal{G}$ will be used in Chapter 4 and in particular in Theorem 4.5. The second pair Q^-, $\mathcal{G}^-$ is applied in Chapter 3, particularly in Theorem 3.7.

Theorem 1.11.
The category $\mathcal{C}_n$ is isomorphic to the category $\mathcal{Q}ui_n^{\Sigma_1}$ using one of the following pairs of (covariant) functors:

(i) Use the functors $Q\colon \mathcal{Q}ui_n^{\Sigma_1} \to \mathcal{C}_n$ and $\mathcal{G}\colon \mathcal{C}_n \to \mathcal{Q}ui_n^{\Sigma_1}$ defined as follows:
Let $(V_I, u_{I,i}, c_{I,i})$ denote an object in $\mathcal{Q}ui_n^{\Sigma_1}$ and let (h_I) denote a morphism in $\mathcal{Q}ui_n^{\Sigma_1}$. We set

$$Q((V_I, u_{I,i}, c_{I,i})) := (V_I, u_{I,i}, y_{I,i}) \quad \text{and} \quad Q((h_I)) := (h_I)$$
$$\text{where } y_{I,i} := \sum_{k=1}^{\infty} \frac{(2\pi i)^k}{k!} (c_{I,i} \circ u_{I,i})^{k-1} \circ c_{I,i} = c_{I,i} \circ \sum_{k=1}^{\infty} \frac{(2\pi i)^k}{k!} (u_{I,i} \circ c_{I,i})^{k-1}.$$

Let $(V_I, u_{I,i}, w_{I,i})$ denote an object in $\mathcal{C}_n$ and let (h_I) denote a morphism in $\mathcal{C}_n$. Then, set

$$\mathcal{G}((V_I, u_{I,i}, w_{I,i})) := (V_I, u_{I,i}, x_{I,i}) \quad \text{and} \quad \mathcal{G}((h_I)) := (h_I).$$

The map $x_{I,i}$ is given as follows: Let $s_{I,i}\colon V_I \to V_I$ denote the unique linear map with eigenvalues in Σ_1 such that

$$e^{2\pi i s_{I,i}} = w_{I,i} \circ u_{I,i} + \mathrm{Id}_{V_I} \quad \text{and set} \quad x_{I,i} := \left(\sum_{k=1}^{\infty} \frac{(2\pi i)^k}{k!} s_{I,i}^{k-1} \right)^{-1} \circ w_{I,i}.$$

(ii) Alternatively, use the similarly defined functors $\mathcal{Q}^-\colon \mathcal{Q}ui_n^{\Sigma_1} \to \mathcal{C}_n$ and $\mathcal{G}^-\colon \mathcal{C}_n \to \mathcal{Q}ui_n^{\Sigma_1}$: Let $(V_I, u_{I,i}, c_{I,i})$ denote an object in $\mathcal{Q}ui_n^{\Sigma_1}$ and let (h_I) denote a morphism in $\mathcal{Q}ui_n^{\Sigma_1}$. We set

$$\mathcal{Q}^-((V_I, u_{I,i}, c_{I,i})) := (V_I, u_{I,i}, y_{I,i}) \quad \textit{and} \quad \mathcal{Q}^-((h_I)) := (h_I)$$
$$\textit{where } y_{I,i} := \sum_{k=1}^{\infty} \frac{(-2\pi i)^k}{k!} (c_{I,i} \circ u_{I,i})^{k-1} \circ c_{I,i} = c_{I,i} \circ \sum_{k=1}^{\infty} \frac{(-2\pi i)^k}{k!} (u_{I,i} \circ c_{I,i})^{k-1}.$$

Let $(V_I, u_{I,i}, w_{I,i})$ denote an object in $\mathcal{C}_n$ and let (h_I) denote a morphism in $\mathcal{C}_n$. Then, let

$$\mathcal{G}^-((V_I, u_{I,i}, w_{I,i})) := (V_I, u_{I,i}, x_{I,i}) \quad \textit{and} \quad \mathcal{G}^-((h_I)) := (h_I).$$

The map $x_{I,i}$ is given as follows: Let $s_{I,i}\colon V_I \to V_I$ denote the unique linear map with eigenvalues in Σ_1 such that

$$e^{-2\pi i s_{I,i}} = w_{I,i} \circ u_{I,i} + \mathrm{Id}_{V_I} \quad \textit{and set} \quad x_{I,i} := \left(\sum_{k=1}^{\infty} \frac{(-2\pi i)^k}{k!} s_{I,i}^{k-1} \right)^{-1} \circ w_{I,i}.$$

Note that i in $2\pi i$ is the imaginary unit. To prevent confusion, we already note that a major part of the proof of Theorem 1.11 is to show that the functors $\mathcal{Q}$, $\mathcal{G}$, respectively $\mathcal{Q}^-$, $\mathcal{G}^-$, are well-defined. However, before proving the theorem, we verify two helpful statements from matrix analysis. The following proposition states the existence and uniqueness of a matrix logarithm once one restricts the eigenvalues.

Proposition 1.12.
Let E denote a finite dimensional $\mathbb{C}$-vector space and let $f\colon E \to E$ denote an invertible linear map. Then there exist unique linear maps $g_1, g_2\colon E \to E$ whose spectrums are contained in Σ_1 such that

$$f = e^{2\pi i g_1} \quad \textit{and} \quad f = e^{-2\pi i g_2}.$$

Proof:
For g_1 we choose the branch of the logarithm defined on $\mathbb{C} \setminus \mathbb{R}_{\geq 0}$ with image contained in $\{\alpha \in \mathbb{C} \mid 0 < \mathrm{Im}(\alpha) < 2\pi\}$, and we use a unique extension to $\mathbb{C} \setminus \{0\}$ with image in $2\pi i \Sigma_1$. For g_2 we choose the branch of the logarithm defined on $\mathbb{C} \setminus \mathbb{R}_{\geq 0}$ with image contained in $\{\alpha \in \mathbb{C} \mid -2\pi < \mathrm{Im}(\alpha) < 0\}$, and we use a unique extension to $\mathbb{C} \setminus \{0\}$ with image in $-2\pi i \Sigma_1$. Whenever the eigenvalues of g_1 and g_2 are contained in Σ_1, the eigenvalues of $2\pi i g_1$ and $-2\pi i g_2$ are contained in the strips

$$2\pi i \Sigma_1 = \left\{ \alpha \in \mathbb{C} \;\middle|\; 0 \leq \mathrm{Im}(\alpha) \leq 2\pi,\ \mathrm{Re}(\alpha) = \begin{cases} \leq 0 & \text{if } \mathrm{Im}(\alpha) = 0, \\ > 0 & \text{if } \mathrm{Im}(\alpha) = 2\pi, \\ \text{arbitrary} & \text{otherwise} \end{cases} \right\}$$

and

$$-2\pi i \Sigma_1 = \left\{ \alpha \in \mathbb{C} \;\middle|\; -2\pi \leq \mathrm{Im}(\alpha) \leq 0,\ \mathrm{Re}(\alpha) = \begin{cases} < 0 & \text{if } \mathrm{Im}(\alpha) = -2\pi, \\ \geq 0 & \text{if } \mathrm{Im}(\alpha) = 0, \\ \text{arbitrary} & \text{otherwise,} \end{cases} \right\}$$

respectively. Note, every complex number has a unique representative in these strips up to $2\pi i \mathbb{N}$.

The existence and uniqueness of g_1 and g_2 follows now with the aid of [HJ91, Corollary 6.2.12 (see also Example 6.2.15 and Example 6.2.16)] which deals with finding a matrix A for a given invertible matrix B such that $e^A = B$. □

The next corollary will be auxiliary to prove commutativity of some diagrams later:

Corollary 1.13.
Let E, F denote two finite dimensional $\mathbb{C}$-vector spaces and let $\gamma\colon E \to F$ denote a linear map. Furthermore, let $\alpha\colon E \to E$ and $\beta\colon F \to F$ denote two linear maps whose eigenvalues are contained in Σ_1. Then:

$$\gamma \circ e^{2\pi i\alpha} = e^{2\pi i\beta} \circ \gamma \quad \textit{if and only if} \quad \gamma \circ \alpha = \beta \circ \gamma \quad \textit{if and only if} \quad \gamma \circ e^{-2\pi i\alpha} = e^{-2\pi i\beta} \circ \gamma$$

Proof:
We will only show the first equivalence. The second equivalence is proved similarly.
The direction "$\Leftarrow$" is easily verified. So we are left with proving the direction "$\Rightarrow$". This proof is divided into three parts:

(i) Assume that $\gamma\colon E \to F$ is invertible. Our equation $\gamma \circ e^{2\pi i\alpha} = e^{2\pi i\beta} \circ \gamma$ is now equivalent to

$$e^{2\pi i\beta} = \gamma \circ e^{2\pi i\alpha} \circ \gamma^{-1} = e^{2\pi i\,(\gamma\circ\alpha\circ\gamma^{-1})}.$$

The eigenvalues of β and $\gamma\circ\alpha\circ\gamma^{-1}$ are both contained in Σ_1. Using the uniqueness given in Proposition 1.12, we obtain the claimed equality

$$\beta = \gamma\circ\alpha\circ\gamma^{-1} \iff \gamma\circ\alpha = \beta\circ\gamma.$$

(ii) To prove the general case, we need the following small statement:

α preserves a linear subspace $\tilde{E}$ of E, i.e. $\alpha(\tilde{E}) \subset \tilde{E}$, if and only if $e^{2\pi i\alpha}$ preserves $\tilde{E}$.

Clearly, if the map α preserves a linear subspace of E, then $e^{2\pi i\alpha}$ will preserve this linear subspace, too. To prove "$\Leftarrow$", we use [HJ91, Corollary 6.2.12] to obtain that α is a polynomial in $e^{2\pi i\alpha}$ (the concrete form of the polynomial depends on the map α). Therefore, α preserves a linear subspace of E if $e^{2\pi i\alpha}$ preserves it. This yields the claimed equivalence.

(iii) Now, let us prove the general case. Examining $\gamma \circ e^{2\pi i\alpha} = e^{2\pi i\beta} \circ \gamma$, we see that $e^{2\pi i\beta}$ preserves $\operatorname{im}(\gamma)$ and that $e^{2\pi i\alpha}$ preserves $\ker(\gamma)$. This gives us a well-defined map

$$\overline{e^{2\pi i\alpha}}\colon E/\ker(\gamma) \to E/\ker(\gamma),$$

and every square of the diagram

$$\begin{array}{ccccccc}
E & \twoheadrightarrow & E/\ker(\gamma) & \xrightarrow{\overline{\gamma}} & \operatorname{im}(\gamma) & \hookrightarrow & F \\
\downarrow{\scriptstyle e^{2\pi i\alpha}} & & \downarrow{\scriptstyle \overline{e^{2\pi i\alpha}}} & & \downarrow{\scriptstyle e^{2\pi i\beta}} & & \downarrow{\scriptstyle e^{2\pi i\beta}} \\
E & \twoheadrightarrow & E/\ker(\gamma) & \xrightarrow{\overline{\gamma}} & \operatorname{im}(\gamma) & \hookrightarrow & F
\end{array}$$

commutes. Using part (ii) of the proof, we obtain that $\operatorname{im}(\gamma)$ and $\ker(\gamma)$ are preserved by

β and α, respectively. This induces a well-defined map

$$\overline{\alpha}\colon {}^{E}\!/_{\ker(\gamma)} \to {}^{E}\!/_{\ker(\gamma)}$$

where $\overline{e^{2\pi i\alpha}} = e^{2\pi i\overline{\alpha}}$. Let us consider the diagram

$$\begin{array}{ccccccc}
E & \twoheadrightarrow & {}^{E}\!/_{\ker(\gamma)} & \xrightarrow{\overline{\gamma}} & \operatorname{im}(\gamma) & \hookrightarrow & F \\
\downarrow{\scriptstyle\alpha} & & \downarrow{\scriptstyle\overline{\alpha}} & & \downarrow{\scriptstyle\beta} & & \downarrow{\scriptstyle\beta} \\
E & \twoheadrightarrow & {}^{E}\!/_{\ker(\gamma)} & \xrightarrow{\overline{\gamma}} & \operatorname{im}(\gamma) & \hookrightarrow & F\,.
\end{array}$$

We already proved that the first and the last square of this diagram commute.

We need to show that $\operatorname{Spec}(\overline{\alpha}) \subset \Sigma_1$: Choose a basis of $\ker(\gamma)$ denoted $e_1, \dots, e_p$ and extend it to a basis of E by $e_{p+1}, \dots, e_m$ where $m = \dim_{\mathbb{C}} E$. We see that α can be represented by a matrix of the form

$$\begin{pmatrix} A & * \\ 0 & B \end{pmatrix}$$

where A is a $p \times p$-matrix and B is a $(m-p)\times(m-p)$-matrix. Using $e_{p+1}, \dots, e_m$ as basis for ${}^{E}\!/_{\ker(\gamma)}$, we see that B represents $\overline{\alpha}$ which shows that $\operatorname{Spec}(\overline{\alpha}) \subset \operatorname{Spec}(\alpha) \subset \Sigma_1$.

As $\overline{\gamma}\colon {}^{E}\!/_{\ker(\gamma)} \to \operatorname{im}(\gamma)$ is invertible and $\operatorname{Spec}(\overline{\alpha})$, $\operatorname{Spec}(\beta) \subset \Sigma_1$, the commutativity of

$$\begin{array}{ccc}
{}^{E}\!/_{\ker(\gamma)} & \xrightarrow{\overline{\gamma}} & \operatorname{im}(\gamma) \\
\downarrow{\scriptstyle e^{2\pi i\overline{\alpha}}} & & \downarrow{\scriptstyle e^{2\pi i\beta}} \\
{}^{E}\!/_{\ker(\gamma)} & \xrightarrow{\overline{\gamma}} & \operatorname{im}(\gamma)
\end{array}
\quad \text{implies that} \quad
\begin{array}{ccc}
{}^{E}\!/_{\ker(\gamma)} & \xrightarrow{\overline{\gamma}} & \operatorname{im}(\gamma) \\
\downarrow{\scriptstyle\overline{\alpha}} & & \downarrow{\scriptstyle\beta} \\
{}^{E}\!/_{\ker(\gamma)} & \xrightarrow{\overline{\gamma}} & \operatorname{im}(\gamma)
\end{array}$$

commutes by part (i) of the proof. Hence, the middle square in the above diagram commutes as well and we obtain

$$\gamma \circ \alpha = \beta \circ \gamma$$

as claimed. □

Now, we are ready to prove Theorem 1.11:

Proof of Theorem 1.11:
We will only demonstrate the proof for the first pair Q and $\mathcal{G}$. The proof for the pair Q^- and $\mathcal{G}^-$ is completely identical to this one (except of several $-2\pi i$ instead of $2\pi i$).

For simplicity, we set

$$\psi(A) := \sum_{k=1}^{\infty} \frac{(2\pi i)^k}{k!} A^{k-1}$$

for a linear map A.

We need to check several small statements to obtain the theorem:

(i) Verify that $x_{I,i}$ is well-defined:
The map $w_{I,i} \circ u_{I,i} + \mathrm{Id}_{V_I}$ is invertible by assumption. Using Proposition 1.12, we obtain the existence and uniqueness of $s_{I,i}$. Furthermore, it is easy to verify that $\psi(s_{I,i})$ is invertible: The eigenvalues of $\psi(s_{I,i})$ are uniquely determined by $s_{I,i}$ in such a way that, if λ is an eigenvalue of $s_{I,i}$, then $\psi(\lambda)$ is an eigenvalue of $\psi(s_{I,i})$. If $\lambda = 0$ then $\psi(0) = 2\pi i \neq 0$. If $\lambda \neq 0$ we may write
$$\psi(\lambda) = \frac{e^{2\pi i\lambda} - 1}{\lambda}.$$
Since $\Sigma_1 \cap \mathbb{Z} = \{0\}$, we see that $\psi(\lambda) \neq 0$ for $\lambda \in \Sigma_1$. This proves that the eigenvalues of $\psi(s_{I,i})$ are non-zero and $x_{I,i} = \psi(s_{I,i})^{-1} \circ w_{I,i}$ is well-defined.

(ii) We have to check that $Q\colon \mathcal{Qui}_n^{\Sigma_1} \to \mathcal{C}_n$ is a well-defined functor:
Let $(V_I, u_{I,i}, c_{I,i})$ denote an object in $\mathcal{Qui}_n^{\Sigma_1}$ and let $(V_I, u_{I,i}, y_{I,i})$ denote its image under Q. $(V_I, u_{I,i}, y_{I,i})$ is indeed an object in $\mathcal{C}_n$:

- The map
$$y_{I,i} \circ u_{I,i} + \mathrm{Id}_{V_I} = \psi(c_{I,i} \circ u_{I,i}) \circ c_{I,i} \circ u_{I,i} + \mathrm{Id}_{V_I} = e^{2\pi i (c_{I,i} \circ u_{I,i})}$$
is obviously invertible.
- The commutativity conditions in $\mathcal{C}_n$ follow easily from those in $\mathcal{Qui}_n^{\Sigma_1}$:

$\star\ u_{I\cup\{i\},j} \circ u_{I,i} = u_{I\cup\{j\},i} \circ u_{I,j}$

$\star\ y_{I,i} \circ y_{I\cup\{i\},j} = c_{I,i} \circ \psi(u_{I,i} \circ c_{I,i}) \circ \psi(c_{I\cup\{i\},j} \circ u_{I\cup\{i\},j}) \circ c_{I\cup\{i\},j} =$
$= c_{I,i} \circ \psi(c_{I\cup\{i\},j} \circ u_{I\cup\{i\},j}) \circ \psi(u_{I,i} \circ c_{I,i}) \circ c_{I\cup\{i\},j} =$
$= c_{I,j} \circ \psi(u_{I,j} \circ c_{I,i}) \circ \psi(c_{I\cup\{i\},j} \circ u_{I\cup\{j\},i}) \circ c_{I\cup\{j\},i} =$
$= c_{I,j} \circ \psi(u_{I,j} \circ c_{I,j}) \circ \psi(c_{I\cup\{j\},i} \circ u_{I\cup\{j\},i}) \circ c_{I\cup\{j\},i} =$
$= y_{I,j} \circ y_{I\cup\{j\},i}$

$\star\ y_{I\cup\{i\},j} \circ u_{I\cup\{j\},i} = c_{I\cup\{i\},j} \circ \psi(u_{I\cup\{i\},j} \circ c_{I\cup\{i\},j}) \circ u_{I\cup\{j\},i} =$
$= c_{I\cup\{i\},j} \circ \psi(u_{I\cup\{i\},j} \circ u_{I,i}) \circ c_{I,j} = u_{I,i} \circ \psi(c_{I,j} \circ u_{I,j}) \circ c_{I,j} =$
$= u_{I,i} \circ y_{I,j}$

Now, let (h_I) denote a morphism from $(V_I, u_{I,i}, c_{I,i})$ to $(V'_I, u'_{I,i}, c'_{I,i})$ in $\mathcal{Qui}_n^{\Sigma_1}$. We have to prove that $Q((h_I)) = (h_I)$ is a morphism from $(V_I, u_{I,i}, y_{I,i})$ to $(V'_I, u'_{I,i}, y'_{I,i})$ in $\mathcal{C}_n$. For this, we need to check the following two equations:

$\star\ u'_{I,i} \circ h_I = h_{I\cup\{i\}} \circ u_{I,i}$

$\star\ h_I \circ y_{I,i} = y'_{I,i} \circ h_{I\cup\{i\}} \iff h_I \circ c_{I,i} \circ \psi(u_{I,i} \circ c_{I,i}) = c'_{I,i} \circ \psi(u'_{I,i} \circ c'_{I,i}) \circ h_{I\cup\{i\}}$

The first one is clearly correct as (h_I) is a morphism in $\mathcal{Qui}_n^{\Sigma_1}$. The second equation follows easily from the first equation and the equality
$$h_I \circ c_{I,i} = c'_{I,i} \circ h_{I\cup\{i\}}\,.$$
So (h_I) is a morphism in $\mathcal{C}_n$, too.

(iii) We have to check that $\mathcal{G}\colon \mathcal{C}_n \to \mathcal{Qui}_n^{\Sigma_1}$ is a well-defined functor:
Let $(V_I, u_{I,i}, w_{I,i})$ denote an object in $\mathcal{C}_n$ and let $(V_I, u_{I,i}, x_{I,i})$ denote its image under $\mathcal{G}$.

$(V_I, u_{I,i}, x_{I,i})$ is indeed an object in $\mathcal{Q}ui_n^{\Sigma_1}$:

- We have the equality

$$x_{I,i} \circ u_{I,i} = \psi(s_{I,i})^{-1} \circ w_{I,i} \circ u_{I,i} = \psi(s_{I,i})^{-1} \circ (e^{2\pi i s_{I,i}} - \mathrm{Id}_{V_I}) = \\ = \psi(s_{I,i})^{-1} \circ \psi(s_{I,i}) \circ s_{I,i} = s_{I,i}\,.$$

  The eigenvalues of $s_{I,i}$ are contained in Σ_1. Thus, those of $x_{I,i} \circ u_{I,i}$ are contained in Σ_1 as well.

- To prove the commutativity conditions we need to check the following identities:

$$\star\ u_{I\cup\{i\},j} \circ u_{I,i} = u_{I\cup\{j\},i} \circ u_{I,j}$$

$$\star\ x_{I,i} \circ x_{I\cup\{i\},j} = x_{I,j} \circ x_{I\cup\{j\},i} \Longleftrightarrow \psi(s_{I,i})^{-1} \circ w_{I,i} \circ \psi(s_{I\cup\{i\},j})^{-1} \circ w_{I\cup\{i\},j} = \\ = \psi(s_{I,j})^{-1} \circ w_{I,j} \circ \psi(s_{I\cup\{j\},i})^{-1} \circ w_{I\cup\{j\},i}$$

$$\star\ x_{I\cup\{i\},j} \circ u_{I\cup\{j\},i} = u_{I,i} \circ x_{I,j} \\ \Longleftrightarrow \psi(s_{I\cup\{i\},j})^{-1} \circ w_{I\cup\{i\},j} \circ u_{I\cup\{j\},i} = u_{I,i} \circ \psi(s_{I,j})^{-1} \circ w_{I,j} \\ \Longleftrightarrow \psi(s_{I\cup\{i\},j})^{-1} \circ u_{I,i} \circ w_{I,j} = u_{I,i} \circ \psi(s_{I,j})^{-1} \circ w_{I,j}$$

  The first equality follows directly. Moreover, we have the following identities in $\mathcal{C}_n$:

$$\star\ e^{2\pi i s_{I,j}} \circ w_{I,i} = w_{I,j} \circ u_{I,j} \circ w_{I,i} + w_{I,i} = w_{I,j} \circ w_{I\cup\{j\},i} \circ u_{I\cup\{i\},j} + w_{I,i} = \\ = w_{I,i} \circ (w_{I\cup\{i\},j} \circ u_{I\cup\{i\},j} + \mathrm{Id}_{V_{I\cup\{i\}}}) = w_{I,i} \circ e^{2\pi i s_{I\cup\{i\},j}}$$

$$\star\ e^{2\pi i s_{I,i}} \circ e^{2\pi i s_{I,j}} = (w_{I,i} \circ u_{I,i} + \mathrm{Id}_{V_I}) \circ (w_{I,j} \circ u_{I,j} + \mathrm{Id}_{V_I}) = \\ = w_{I,i} \circ w_{I\cup\{i\},j} \circ u_{I\cup\{j\},i} \circ u_{I,j} + w_{I,i} \circ u_{I,i} + w_{I,j} \circ u_{I,j} + \mathrm{Id}_{V_I} = \\ = w_{I,j} \circ w_{I\cup\{j\},i} \circ u_{I\cup\{i\},j} \circ u_{I,i} + w_{I,i} \circ u_{I,i} + w_{I,j} \circ u_{I,j} + \mathrm{Id}_{V_I} = \\ = w_{I,j} \circ u_{I,j} \circ w_{I,i} \circ u_{I,i} + w_{I,i} \circ u_{I,i} + w_{I,j} \circ u_{I,j} + \mathrm{Id}_{V_I} = \\ = (w_{I,j} \circ u_{I,j} + \mathrm{Id}_{V_I}) \circ (w_{I,i} \circ u_{I,i} + \mathrm{Id}_{V_I}) = e^{2\pi i s_{I,j}} \circ e^{2\pi i s_{I,i}}$$

$$\star\ u_{I,i} \circ e^{2\pi i s_{I,j}} = u_{I,i} \circ (w_{I,j} \circ u_{I,j} + \mathrm{Id}_{V_I}) = w_{I\cup\{i\},j} \circ u_{I\cup\{j\},i} \circ u_{I,j} + u_{I,i} = \\ = w_{I\cup\{i\},j} \circ u_{I\cup\{i\},j} \circ u_{I,i} + u_{I,i} = e^{2\pi i s_{I\cup\{i\},j}} \circ u_{I,i}$$

  Using Corollary 1.13, we obtain:

$$\star\ s_{I,j} \circ w_{I,i} = w_{I,i} \circ s_{I\cup\{i\},j} \quad \Longrightarrow \quad w_{I,i} \circ \psi(s_{I\cup\{i\},j})^{-1} = \psi(s_{I,j})^{-1} \circ w_{I,i}$$

$$\star\ s_{I,i} \circ s_{I,j} = s_{I,j} \circ s_{I,i} \quad \Longrightarrow \quad \psi(s_{I,i})^{-1} \circ \psi(s_{I,j})^{-1} = \psi(s_{I,j})^{-1} \circ \psi(s_{I,i})^{-1}$$

$$\star\ u_{I,i} \circ s_{I,j} = s_{I\cup\{i\},j} \circ u_{I,i} \quad \Longrightarrow \quad \psi(s_{I\cup\{i\},j})^{-1} \circ u_{I,i} = u_{I,i} \circ \psi(s_{I,j})^{-1}$$

  Using these three equations, the remaining two commutativity conditions in $\mathcal{Q}ui_n^{\Sigma_1}$ are easily verified.

Now, let (h_I) denote a morphism from $(V_I, u_{I,i}, w_{I,i})$ to $(V'_I, u'_{I,i}, w'_{I,i})$ in $\mathcal{C}_n$. We have to prove that $\mathcal{G}((h_I)) = (h_I)$ is a morphism from $(V_I, u_{I,i}, x_{I,i})$ to $(V'_I, u'_{I,i}, x'_{I,i})$ in $\mathcal{Q}ui_n^{\Sigma_1}$. Therefor we need to check the following equations:

$$\star\ u'_{I,i} \circ h_I = h_{I\cup\{i\}} \circ u_{I,i}$$

$$\star\ h_I \circ x_{I,i} = x'_{I,i} \circ h_{I\cup\{i\}} \Longleftrightarrow h_I \circ \psi(s_{I,i})^{-1} \circ w_{I,i} = \psi(s'_{I,i})^{-1} \circ w'_{I,i} \circ h_{I\cup\{i\}} \\ \Longleftrightarrow h_I \circ \psi(s_{I,i})^{-1} \circ w_{I,i} = \psi(s'_{I,i})^{-1} \circ h_I \circ w_{I,i}$$

The first one follows directly. To prove the second equation we use the equality

$$e^{2\pi i s'_{I,i}} \circ h_I = (w'_{I,i} \circ u'_{I,i} + \mathrm{Id}_{V'_I}) \circ h_I = w'_{I,i} \circ h_{I\cup\{i\}} \circ u_{I,i} + h_I = \\ = h_I \circ w_{I,i} \circ u_{I,i} + h_I = h_I \circ e^{2\pi i s_{I,i}}$$

in $\mathcal{C}_n$. Then Corollary 1.13 yields

$$s'_{I,i} \circ h_I = h_I \circ s_{I,i} \quad \text{and therefore} \quad h_I \circ \psi(s_{I,i})^{-1} = \psi(s'_{I,i})^{-1} \circ h_I\,.$$

This finally proves the second equation and shows that (h_I) is a morphism in $\mathcal{Q}ui_n^{\Sigma_1}$.

(iv) We show that $\mathcal{Q} \circ \mathcal{G} = \mathrm{Id}_{\mathcal{C}_n}$:
Clearly, $\mathcal{Q}\circ\mathcal{G}$ is the identity on morphisms. So, we need to check for an object $(V_I, u_{I,i}, w_{I,i})$ in $\mathcal{C}_n$ that $(\mathcal{Q} \circ \mathcal{G})((V_I, u_{I,i}, w_{I,i})) = (V_I, u_{I,i}, w_{I,i})$. Let

$$\mathcal{G}((V_I, u_{I,i}, w_{I,i})) =: (V_I, u_{I,i}, x_{I,i}) \quad \text{where} \quad x_{I,i} = \psi(s_{I,i})^{-1} \circ w_{I,i}\,.$$

And let

$$\mathcal{Q}((V_I, u_{I,i}, x_{I,i})) =: (V_I, u_{I,i}, y_{I,i}) \quad \text{where} \quad y_{I,i} = \psi(x_{I,i} \circ u_{I,i}) \circ x_{I,i}\,.$$

By part (iii) of the proof, $x_{I,i}$, $u_{I,i}$ and $s_{I,i}$ fulfil the equality $x_{I,i} \circ u_{I,i} = s_{I,i}$. Using the definition of $x_{I,i}$, this yields

$$y_{I,i} = \psi(x_{I,i} \circ u_{I,i}) \circ x_{I,i} = \psi(s_{I,i}) \circ \psi(s_{I,i})^{-1} \circ w_{I,i} = w_{I,i}\,.$$

All in all, this shows that $\mathcal{Q} \circ \mathcal{G}$ is the identity on objects, too.

(iv) We show that $\mathcal{G} \circ \mathcal{Q} = \mathrm{Id}_{\mathcal{Q}ui_n^{\Sigma_1}}$:
Clearly, $\mathcal{G}\circ\mathcal{Q}$ is the identity on morphisms. So, we need to verify for an object $(V_I, u_{I,i}, c_{I,i})$ in $\mathcal{Q}ui_n^{\Sigma_1}$ that $(\mathcal{G} \circ \mathcal{Q})((V_I, u_{I,i}, c_{I,i})) = (V_I, u_{I,i}, c_{I,i})$. We set

$$\mathcal{Q}((V_I, u_{I,i}, c_{I,i})) =: (V_I, u_{I,i}, y_{I,i}) \quad \text{where} \quad y_{I,i} = \psi(c_{I,i} \circ u_{I,i}) \circ c_{I,i}\,.$$

And let

$$\mathcal{G}((V_I, u_{I,i}, y_{I,i})) =: (V_I, u_{I,i}, x_{I,i}) \quad \text{where} \quad x_{I,i} = \psi(s_{I,i})^{-1} \circ y_{I,i}\,.$$

We have the equality

$$e^{2\pi i (c_{I,i} \circ u_{I,i})} = y_{I,i} \circ u_{I,i} + \mathrm{Id}_{V_I} = e^{2\pi i s_{I,i}}\,.$$

The eigenvalues of $c_{I,i} \circ u_{I,i}$ and $s_{I,i}$ are contained in Σ_1. Thus, the uniqueness of $s_{I,i}$ (cf. Proposition 1.12) yields $c_{I,i} \circ u_{I,i} = s_{I,i}$. Hence,

$$x_{I,i} = \psi(s_{I,i})^{-1} \circ y_{I,i} = \psi(c_{I,i} \circ u_{I,i})^{-1} \circ \psi(c_{I,i} \circ u_{I,i}) \circ c_{I,i} = c_{I,i}\,.$$

All in all, this shows that $\mathcal{G} \circ \mathcal{Q}$ is the identity on objects, too.

This shows that $\mathcal{Q}\colon \mathcal{Q}ui_n^{\Sigma_1} \to \mathcal{C}_n$ and $\mathcal{G}\colon \mathcal{C}_n \to \mathcal{Q}ui_n^{\Sigma_1}$ are inverse functors to each other and therefore they define an isomorphism between the categories $\mathcal{Q}ui_n^{\Sigma_1}$ and $\mathcal{C}_n$. □

Part II.

Regular singular $\mathscr{D}$-modules in $\mathbb{C}^n$ whose singular locus is a normal crossing

2. Quiver $\mathscr{D}$-modules in $\mathbb{C}^n$ whose singular locus is a normal crossing

From now on $\mathcal{O} = \mathcal{O}_X$ always denotes the sheaf of analytic functions on $X = \mathbb{C}^n$ for a fixed integer $n \in \mathbb{N}^+$, and $\mathscr{D} = \mathscr{D}_X$ denotes the sheaf of rings of linear partial differential operators with analytic coefficients. Furthermore, we denote by $z_1, \ldots, z_n$ the coordinates of $\mathbb{C}^n$ and by $\partial_i = \frac{\partial}{\partial z_i}$ the i-th partial derivation operator for $i = 1, \ldots, n$.

In this chapter we are going to consider quiver $\mathscr{D}$-modules. These are $\mathscr{D}$-modules which are defined on the basis of certain representations of quivers. We are going to use the quiver representation category $\mathcal{Q}ui_n$ as starting point for our quiver $\mathscr{D}$-modules. Quiver $\mathscr{D}$-modules are highly interesting as the category $\mathcal{M}od^S_{\mathrm{rh}}(\mathscr{D})$ of regular singular holonomic $\mathscr{D}$-modules on $\mathbb{C}^n$ whose singular locus is a normal crossing (see Definition 2.3) is equivalent to the category $\mathcal{C}_n$. This is explained in Theorem 4.2 and thereafter. It means that every such $\mathscr{D}$-module is uniquely determined by a hypercube quiver representation up to an isomorphism.

2.1. Definitions and basic properties

In this section we are first going to define objects of quiver $\mathscr{D}$-modules. This definition (Definition 2.1) is very natural. In Theorem 2.4 we will see that this construction yields a functor E from the quiver representation category $\mathcal{Q}ui_n$ into the $\mathscr{D}$-module category $\mathcal{M}od^S_{\mathrm{rh}}(\mathscr{D})$. The category of quiver $\mathscr{D}$-modules will be interpreted as the essential image of this functor. A main goal of the present work is to show that this category of quiver $\mathscr{D}$-modules is equivalent to the category $\mathcal{M}od^S_{\mathrm{rh}}(\mathscr{D})$. Hereto, we are going to prove in Chapter 4 that, loosely speaking, E is almost a quasi-inverse of the above-mentioned functor from $\mathcal{M}od^S_{\mathrm{rh}}(\mathscr{D})$ into $\mathcal{C}_n$ (recall that $\mathcal{C}_n$ and $\mathcal{Q}ui_n^{\Sigma_1}$ are isomorphic categories by Theorem 1.11).

The following construction of a quiver $\mathscr{D}$-module is based on the paper [KV06]. Khoroshkin and Varchenko consider in this paper $\mathscr{D}$-modules associated to given representations of quivers. These quivers are obtained from arrangements of hyperplanes and the induced natural stratification of $\mathbb{C}^n$. This stratification determines the characteristic variety of the quiver $\mathscr{D}$-module. Combining Section 2.2 with Theorem 2.4, we are going to reprove this statement in our case. Khoroshkin and Varchenko are not only considering normal crossings, i. e. they are starting with other quivers besides the hypercube quiver from Chapter 1. We note that their construction in [KV06] is a generalization of an ancient construction of quiver $\mathscr{D}$-modules by Khoroshkin in [Kho95] and [KS98]. The explicit computations which show the connection between our Definition 2.1 and their definition in [KV06] may be found in Section 2.2.

In [KV06] they are able to obtain plenty of small general results and some major results e. g. on a free resolution of a quiver $\mathscr{D}$-module. However, a general placement of quiver $\mathscr{D}$-modules into the context of $\mathscr{D}$-modules is missing. Their statement on the equivalence of categories of quiver $\mathscr{D}$-modules (in the case of a central arrangement) leaves open many questions, in particular on the essential image – see also Remark 4.9. Furthermore, in general they work with algebraic $\mathscr{D}$-modules instead of analytic ones.

Now, let us define the (left) $\mathscr{D}$-module associated to an object from $\mathcal{Q}ui_n$:

Definition 2.1 (Variant of [KV06])
Let $\mathcal{V}_n = (V_I, B_{I\cup\{i\},I}, B_{I,I\cup\{i\}})$ with $I \in \mathcal{P}(\{1,\ldots,n\})$ denote an object of the category $\mathcal{Q}ui_n$ (see Definition 1.1). We define the associated QUIVER $\mathscr{D}$-MODULE $E\mathcal{V}_n$ *as the quotient of*

$$\bigoplus_{I\in\mathcal{P}(\{1,\ldots,n\})} \left(\mathscr{D} \otimes_{\mathbb{C}} \overline{\Omega}_I \otimes_{\mathbb{C}} V_I\right)$$

by the subsheaf $\mathcal{J}$. The sections of $\mathcal{J}$ over $U \subset \mathbb{C}^n$, open, are given by $\mathbb{C}$-linear combinations of the following elements

$$a\partial_i \otimes_{\mathbb{C}} \omega_I \otimes_{\mathbb{C}} v_I - a \otimes_{\mathbb{C}} \omega_{I\cup\{i\}} \otimes_{\mathbb{C}} B_{I\cup\{i\},I}(v_I) \quad and$$
$$az_i \otimes_{\mathbb{C}} \omega_{I\cup\{i\}} \otimes_{\mathbb{C}} v_{I\cup\{i\}} - a \otimes_{\mathbb{C}} \omega_I \otimes_{\mathbb{C}} B_{I,I\cup\{i\}}(v_{I\cup\{i\}})$$

where $I \neq \{1,\ldots,n\}$, $i \in \{1,\ldots,n\} \setminus I$, $a \in \mathscr{D}(U)$, $v_J \in V_J$ for $J \in \mathcal{P}(\{1,\ldots,n\})$ and $\overline{\Omega}_J := \{c\omega_J \mid c \in \mathbb{C}\}$ is a 1-dimensional $\mathbb{C}$-vector space generated by the element ω_J.
The left $\mathscr{D}$-module structure on $E\mathcal{V}_n$ is given by left multiplication.

$\overline{\Omega}_J$ is used here to simplify the notation or to clarify to which summand of $\bigoplus_I \left(\mathscr{D} \otimes_{\mathbb{C}} \overline{\Omega}_I \otimes_{\mathbb{C}} V_I\right)$ an element belongs to.

The following graphic illustrates the way the subsheaf $\mathcal{J}$ is build out of the quiver representation:

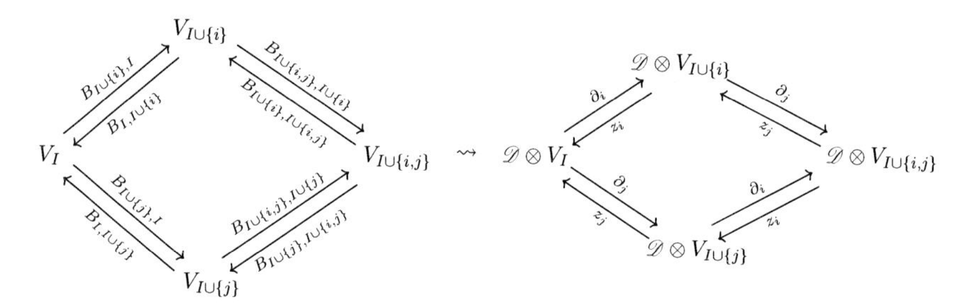

Our next aim is to obtain a functor from $\mathcal{Q}ui_n$ into the category of $\mathscr{D}$-modules on $\mathbb{C}^n$. The following corollary will show us that Definition 2.1 can be extended to the level of morphisms:

Corollary 2.1.
Let $\mathcal{V}_n = (V_I, B_{I\cup\{i\},I}, B_{I,I\cup\{i\}})$ and $\mathcal{V}'_n = (V'_I, B'_{I\cup\{i\},I}, B'_{I,I\cup\{i\}})$ denote two objects in $\mathcal{Q}ui_n$ and let

$$\phi = (h_I)\colon \mathcal{V}_n \to \mathcal{V}'_n$$

denote a morphism from $\mathcal{V}_n$ to $\mathcal{V}'_n$. Then ϕ induces in a natural way a morphism from $E\mathcal{V}_n$ to $E\mathcal{V}'_n$, denoted $E\phi$,

$$E\phi\colon E\mathcal{V}_n \to E\mathcal{V}'_n\,.$$

Proof:
Consider the following diagram:

$$\begin{array}{ccccccccc}
0 \longrightarrow & \mathcal{J} & \hookrightarrow & \bigoplus_{I\in\mathcal{P}(\{1,\dots,n\})} (\mathscr{D}\otimes_{\mathbb{C}}\overline{\Omega}_I\otimes_{\mathbb{C}} V_I) & \twoheadrightarrow & E\mathcal{V}_n & \longrightarrow 0 \\
& \downarrow \tilde{h} & \circlearrowleft & \downarrow \bigoplus_I(\mathrm{Id}_{\mathscr{D}}\otimes\mathrm{Id}_{\overline{\Omega}_I}\otimes h_I) & & \vdots & \\
0 \longrightarrow & \mathcal{J}' & \hookrightarrow & \bigoplus_{I\in\mathcal{P}(\{1,\dots,n\})} (\mathscr{D}\otimes_{\mathbb{C}}\overline{\Omega}_I\otimes_{\mathbb{C}} V'_I) & \twoheadrightarrow & E\mathcal{V}'_n & \longrightarrow 0
\end{array}$$

The rows are exact sequences. Moreover, (h_I) induces naturally a $\mathscr{D}$-linear map $\tilde{h}\colon \mathcal{J} \to \mathcal{J}'$ which fulfils on sections over $U \subset \mathbb{C}^n$, open,

$$\begin{aligned}
&\tilde{h}\big(a\partial_i\otimes\omega_I\otimes v_I - a\otimes\omega_{I\cup\{i\}}\otimes B_{I\cup\{i\},I}(v_I)\big) = \\
&\quad = a\partial_i\otimes\omega_I\otimes h_I(v_I) - a\otimes\omega_{I\cup\{I\}}\otimes h_{I\cup\{i\}}(B_{I\cup\{i\},I}(v_I)) = \\
&\quad = a\partial_i\otimes\omega_I\otimes h_I(v_I) - a\otimes\omega_{I\cup\{I\}}\otimes B'_{I\cup\{i\},I}(h_I(v_I)) \qquad \text{and} \\
&\tilde{h}\big(az_i\otimes\omega_{I\cup\{i\}}\otimes v_{I\cup\{i\}} - a\otimes\omega_I\otimes B_{I,I\cup\{i\}}(v_{I\cup\{i\}})\big) = \\
&\quad = az_i\otimes\omega_{I\cup\{i\}}\otimes h_{I\cup\{i\}}(v_{I\cup\{i\}}) - a\otimes\omega_I\otimes h_I(B_{I,I\cup\{i\}}(v_{I\cup\{i\}})) = \\
&\quad = az_i\otimes\omega_{I\cup\{i\}}\otimes h_{I\cup\{i\}}(v_{I\cup\{i\}}) - a\otimes\omega_I\otimes B'_{I,I\cup\{i\}}(h_{I\cup\{i\}}(v_{I\cup\{i\}}))\,.
\end{aligned}$$

This makes the square commute as indicated. In particular, it induces in a natural way a $\mathscr{D}$-linear morphism from $E\mathcal{V}_n$ to $E\mathcal{V}'_n$. □

Definition 2.1 and Corollary 2.1 altogether give us a functor from the category $\mathcal{Q}ui_n$ into the category of $\mathscr{D}$-modules on $\mathbb{C}^n$:

Proposition/Definition 2.2.
Let $\mathcal{Q}ui_n$ denote the category of finite hypercube quiver representations (see Definition 1.1) and let $\mathcal{M}od(\mathscr{D})$ denote the category of $\mathscr{D}$-modules on $\mathbb{C}^n$. Then we obtain a (covariant) functor, denoted E, from the category $\mathcal{Q}ui_n$ into the category of $\mathscr{D}$-modules

$$E\colon \mathcal{Q}ui_n \to \mathcal{M}od(\mathscr{D})\,.$$

E associates to an object $\mathcal{V}_n$ in $\mathcal{Q}ui_n$ the object $E\mathcal{V}_n$ in $\mathcal{M}od(\mathscr{D})$ (see Definition 2.1), and it associates to a morphism $\phi\colon \mathcal{V}_n \to \mathcal{V}'_n$ in $\mathcal{Q}ui_n$ the morphism $E\phi\colon E\mathcal{V}_n \to E\mathcal{V}'_n$ in $\mathcal{M}od(\mathscr{D})$ (see Corollary 2.1). The category of quiver $\mathscr{D}$-modules is the essential image of the functor E.

Proof:
Let us use Corollary 2.1: We immediately see that E preserves the identity morphism. Furthermore, let (h_I) and (g_I) denote two morphisms in $\mathcal{Q}ui_n$ with compatible source and target, respectively. The fact that

$$(\mathrm{Id}_{\mathscr{D}}\otimes\mathrm{Id}_{\overline{\Omega}_I}\otimes h_I)\circ(\mathrm{Id}_{\mathscr{D}}\otimes\mathrm{Id}_{\overline{\Omega}_I}\otimes g_I) = (\mathrm{Id}_{\mathscr{D}}\otimes\mathrm{Id}_{\overline{\Omega}_I}\otimes(h_I\circ g_I)) \quad\text{and}\quad \widetilde{h\circ g} = \tilde{h}\circ\tilde{g}$$

shows us that E preserves the composition of morphisms, too. Hence, E is indeed a functor from $\mathcal{Q}ui_n$ to $\mathcal{M}od(\mathscr{D})$. □

Now, let us define the category $\mathcal{M}od^S_{\mathrm{rh}}(\mathscr{D})$ of regular singular holonomic $\mathscr{D}$-modules whose singular locus is (contained in) the normal crossing. However, we have to mention that this denomination is a little bit sloppy as in fact the objects in $\mathcal{M}od^S_{\mathrm{rh}}(\mathscr{D})$ needs to fulfil a property on their characteristic variety from which the property on their singular locus follows.

Definition 2.3.
Let $S := \{z_1 \cdot \ldots \cdot z_n = 0\}$ denote the normal crossing in $\mathbb{C}^n$. S induces naturally a (Whitney) stratification of $\mathbb{C}^n$ by 2^n disjoint strata $X_I \subset \mathbb{C}^n$ which are defined by

$$\overline{X}_I := \{z_i = 0 \mid i \in I\} \quad and \quad X_I := \overline{X}_I \setminus \left(\bigcup_{\substack{J \in \mathcal{P}(\{1,\ldots,n\}) \\ \overline{X}_J \subsetneq \overline{X}_I}} \overline{X}_J \right)$$

for $I \in \mathcal{P}(\{1,\ldots,n\})$. This fulfils $X_\varnothing = \mathbb{C}^n \setminus S$ and $S = \bigcup_{I, \dim X_I < n} X_I$.
The category $\mathcal{M}od^S_{rh}(\mathscr{D})$ is then defined to be the category of regular singular holonomic $\mathscr{D}$-modules whose characteristic variety is contained in

$$\bigcup_{I \in \mathcal{P}(\{1,\ldots,n\})} T^*_{X_I}\mathbb{C}^n$$

where

$$\begin{aligned} T^*_{X_I}\mathbb{C}^n &= \{(z,\xi) = (z_1,\ldots,z_n,\xi_1,\ldots,\xi_n) \in T^*\mathbb{C}^n \mid z \in X_I,\ \xi(v) = 0 \ for\ all\ v \in TX_I\} = \\ &= \{(z_1,\ldots,z_n,\xi_1,\ldots,\xi_n) \in T^*\mathbb{C}^n \mid z_i = 0 \Leftrightarrow i \in I,\ \xi_i = 0\ for\ i \notin I\}\,. \end{aligned}$$

We note the following fact which simplifies Definition 2.3 a little bit:
Let

$$\Delta_S := \{(z_1,\ldots,z_n,\xi_1,\ldots,\xi_n) \in T^*\mathbb{C}^n \mid z_i\xi_i = 0 \text{ for all } i \in \{1,\ldots,n\}\}\,.$$

Then it is easy to see that

$$\bigcup_{I \in \mathcal{P}(\{1,\ldots,n\}} T^*_{X_I}\mathbb{C}^n = \Delta_S\,.$$

Therefore, $\mathcal{M}od^S_{\mathrm{rh}}(\mathscr{D})$ is exactly the category of regular singular holonomic $\mathscr{D}$-modules whose characteristic variety is contained in Δ_S.

Now, let us explain why the property on the singular locus is implied by this:
Let $\mathscr{M}$ denote a holonomic $\mathscr{D}$-module and let $\mathrm{Char}(\mathscr{M})$ denote its characteristic variety. The singular locus of $\mathscr{M}$ is defined as the projection of $\mathrm{Char}(\mathscr{M}) \setminus T^*_{\mathbb{C}^n}\mathbb{C}^n$ onto $\mathbb{C}^n$ where $T^*_{\mathbb{C}^n}\mathbb{C}^n$ is the zero section of $T^*\mathbb{C}^n$. It is easy to see that the projection of $\Delta_S \setminus T^*_{\mathbb{C}^n}\mathbb{C}^n$ to $\mathbb{C}^n$ is equal to $S = \bigcup_{I, \dim X_I < n} X_I$. Hence, the singular locus of every $\mathscr{D}$-module in $\mathcal{M}od^S_{\mathrm{rh}}(\mathscr{D})$ is contained in the normal crossing S.

So we are ready to show that every quiver $\mathscr{D}$-module is indeed a regular singular holonomic $\mathscr{D}$-module and that its characteristic variety is contained in Δ_S, or more precisely we show that the essential image of E is contained in $\mathcal{M}od^S_{\mathrm{rh}}(\mathscr{D})$:

Theorem 2.4.
The functor E from Proposition/Definition 2.2 is a functor from the category $\mathcal{Q}ui_n$ into the category $\mathcal{M}od^S_{rh}(\mathscr{D})$.

Proof:
Let $\mathcal{V}_n = (V_J, B_{J\cup\{j\},J}, B_{J,J\cup\{j\}})$ denote an object in $\mathcal{Q}ui_n$. To prove that the essential image of E is indeed contained in $\mathcal{M}od^S_{\mathrm{rh}}(\mathscr{D})$, we use the good filtration $F_k E\mathcal{V}_n$ on $E\mathcal{V}_n$. $F_k E\mathcal{V}_n$ is the filtration induced by the exact sequence

$$\mathscr{D} \otimes \left(\bigoplus_J \overline{\Omega}_J \otimes V_J \right) \twoheadrightarrow E\mathcal{V}_n \to 0$$

where we use the standard filtration $F_\bullet \mathscr{D}$ of $\mathscr{D}$. Recall that $F_k\mathscr{D}$ is the subsheaf of $\mathscr{D}$ of differential operators of order at most $k \in \mathbb{Z}$ and $F_k\mathscr{D} = 0$ for $k < 0$.
Let $P \in F_k\mathscr{D}(U)$ for $U \subset \mathbb{C}^n$, open, and $v_I \in V_I$ for $I \in \mathcal{P}(\{1,\dots,n\})$. We denote by $[P \otimes \omega_I \otimes v_I]$ the image of

$$P \otimes \omega_I \otimes v_I \in F_k\mathscr{D}(U) \otimes \overline{\Omega}_I \otimes V_I$$

in $F_k E\mathcal{V}_n$. Consider

$$\mathrm{gr}^F_k E\mathcal{V}_n = F_k E\mathcal{V}_n \big/ F_{k-1} E\mathcal{V}_n$$

and let $\sigma_k[P \otimes \omega_I \otimes v_I]$ be the image of $[P \otimes \omega_I \otimes v_I]$ in $\mathrm{gr}^F_k E\mathcal{V}_n$. The graded ring

$$\mathrm{gr}^F \mathscr{D} = \bigoplus_k \mathrm{gr}^F_k \mathscr{D} \cong \mathcal{O}_{\mathbb{C}^n}[\xi_1,\dots,\xi_n]$$

acts naturally on $\mathrm{gr}^F E\mathcal{V}_n = \bigoplus_k \mathrm{gr}^F_k E\mathcal{V}_n$. We will prove that $z_i\xi_i$ annihilates $\mathrm{gr}^F_k E\mathcal{V}_n$ for every $k \in \mathbb{Z}$ and every $i \in \{1,\dots,n\}$. This shows that the characteristic variety must be contained in Δ_S. We need to distinguish two cases:
If $i \notin I$, then

$$\begin{aligned} z_i\xi_i \cdot \sigma_k[P \otimes \omega_I \otimes v_I] &= \sigma_{k+1}[z_i\partial_i P \otimes \omega_I \otimes v_I] = \sigma_{k+1}[P z_i\partial_i \otimes \omega_I \otimes v_I] = \\ &= \sigma_{k+1}[P \otimes \omega_I \otimes B_{I,I\cup\{i\}} B_{I\cup\{i\},I}(v_I)] = 0\,. \end{aligned}$$

If $i \in I$, then

$$\begin{aligned} z_i\xi_i \cdot \sigma_k[P \otimes \omega_I \otimes v_I] &= \sigma_{k+1}[z_i\partial_i P \otimes \omega_I \otimes v_I] = \sigma_{k+1}[P \partial_i z_i \otimes \omega_I \otimes v_I] = \\ &= \sigma_{k+1}[P \otimes \omega_I \otimes B_{I,I\setminus\{i\}} B_{I\setminus\{i\},I}(v_I)] = 0\,. \end{aligned}$$

We see that in both cases $z_i\xi_i$ is an annihilator and the characteristic variety of $E\mathcal{V}_n$ is contained in Δ_S as claimed.
This also shows us that the dimension of the characteristic variety of $E\mathcal{V}_n$ is at most $n = \dim_{\mathbb{C}} X$ and therefore $E\mathcal{V}_n$ is holonomic. And as well, we immediately see that $E\mathcal{V}_n$ is regular holonomic using the fact that $(z_i\xi_i)^1$ is an annihilator of $\mathrm{gr}^F E\mathcal{V}_n$ (cf. [Kas03, Definition 5.2]). □

In [KV06] a similar but slightly different proof of the holonomicity and the statement on the characteristic variety is given. Khoroshkin and Varchenko are using global sections as their quiver $\mathscr{D}$-module is algebraic and not analytic. The regular singularity is stated in [KS98] but not rigorously proved.

2.2. Connection to the quiver $\mathscr{D}$-modules of Khoroshkin and Varchenko

In this section we show how the Definition 2.1 of a quiver $\mathscr{D}$-module associated to a hypercube quiver representation is related to the definition of Khoroshkin and Varchenko given in [KV06, Subsection 4.2]. We note that in [KV06] a representation of a quiver is called itself a quiver which seems to be inconsistent with the usual nomenclature in the theory of quiver representations. To simplify comparisons between this Section 2.2 and [KV06], we use however the nomenclature of Khoroshkin and Varchenko for a moment. Moreover, we assume that the reader is familiar with the construction of Khoroshkin and Varchenko or will read [KV06] parallel to this section.

Given the hyperplanes $H_i := \{z_i = 0\}$ in $\mathbb{C}^n$ for $i = 1, \ldots, n$, we obtain a decomposition of $\mathbb{C}^n$ into 2^n disjoint strata X_I with $I \in \mathcal{P}(\{1, \ldots, n\})$ by setting

$$\overline{X}_I := \bigcap_{i \in I} H_i \quad \text{and} \quad X_I := \overline{X}_I \setminus \left(\bigcup_{\substack{J \in \mathcal{P}(\{1,\ldots,n\}) \\ \overline{X}_J \subsetneq \overline{X}_I}} \overline{X}_J \right).$$

This is exactly the stratification of $\mathbb{C}^n$ from Definition 2.3. For $i \in \{1, \ldots, n\} \setminus I$ we see that

$$\overline{X}_{I \cup \{i\}} \subset \overline{X}_I$$

and we get a relation in codimension 1 in the sense that

$$\operatorname{codim}_{\overline{X}_I} \overline{X}_{I \cup \{i\}} = 1 .$$

Using this codimension 1 property, we define the relation $\succ$ on $\mathcal{P}(\{1, \ldots, n\})$ by

$$I \succ J \;:\Longleftrightarrow\; \left\{ \overline{X}_I \supset \overline{X}_J, \operatorname{codim}_{\overline{X}_I} \overline{X}_J = 1 \right\} \;\Longleftrightarrow\; J = I \cup \{i\}, i \notin I$$

and we set

$$l(I) := \operatorname{codim}_{\mathbb{C}^n} \overline{X}_I = |I| .$$

We obtain an unoriented graph as follows:

$$I \text{ --- } I \cup \{i\}$$

The vertices are given by the 2^n elements of $\mathcal{P}(\{1, \ldots, n\})$, and an edge between I and J if and only if $I \succ J$ or $J \succ I$. This yields $n2^{n-1}$ edges.

The quiver $\mathcal{U}_n$ corresponding to this hyperplane arrangement is defined as follows:
Replace the 2^n vertices of the unoriented graph by finite dimensional $\mathbb{C}$-vector spaces and replace each edge by two linear maps in the following way:

$$\mathcal{U}_n = \left(V_I \underset{A_{I, I \cup \{i\}}}{\overset{A_{I \cup \{i\}, I}}{\rightleftarrows}} V_{I \cup \{i\}} \right)$$

We assume this quiver $\mathcal{U}_n$ to be anti-commuting

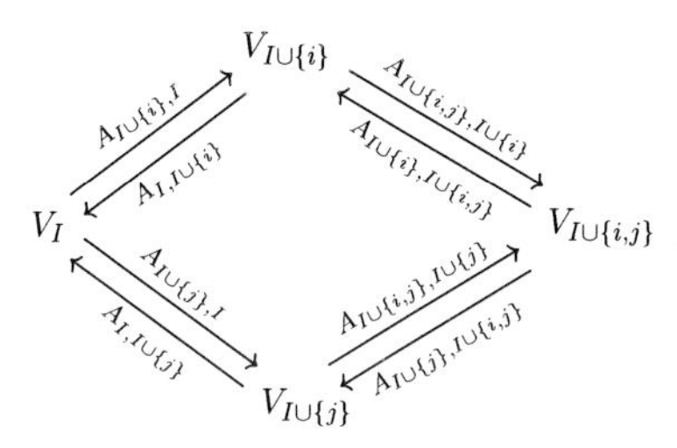

which means that

$$\begin{aligned} A_{I\cup\{i,j\},I\cup\{i\}} \circ A_{I\cup\{i\},I} &= -A_{I\cup\{i,j\},I\cup\{j\}} \circ A_{I\cup\{j\},I}\,, \\ A_{I,I\cup\{i\}} \circ A_{I\cup\{i\},I\cup\{i,j\}} &= -A_{I,I\cup\{j\}} \circ A_{I\cup\{j\},I\cup\{i,j\}}\,, \\ A_{I\cup\{j\},I} \circ A_{I,I\cup\{i\}} &= -A_{I\cup\{j\},I\cup\{i,j\}} \circ A_{I\cup\{i,j\},I\cup\{i\}} \quad \text{and} \\ A_{I\cup\{i\},I} \circ A_{I,I\cup\{j\}} &= -A_{I\cup\{i\},I\cup\{i,j\}} \circ A_{I\cup\{i,j\},I\cup\{j\}}\,. \end{aligned}$$

Obviously, this quiver is not an object in our category $\mathcal{Q}ui_n$. Later we will redefine it in such a way that it is commuting but the corresponding quiver $\mathscr{D}$-module will remain the same.

Before defining the $\mathscr{D}$-module that corresponds to $\mathcal{U}_n$, we need to clarify some terms which only depend on the strata X_I but not on $\mathcal{U}_n$:

- Let $G := \{g(z_1,\ldots,z_n) = a_0 + \sum_{l=1}^n a_l z_l \mid a_l \in \mathbb{C}\}$ denote the space of all affine functions which map from $\mathbb{C}^n$ to $\mathbb{C}$. For $I \in \mathcal{P}(\{1,\ldots,n\})$ we denote by $G_I \subset G$ the space of affine functions vanishing on X_I. We have

$$G_I = \Big\{ g(z_1,\ldots,z_n) = \sum_{\substack{l=1\\ l\in I}}^{n} a_l z_l \mid a_l \in \mathbb{C} \Big\}.$$

- We denote by T the space of all vector fields on $\mathbb{C}^n$ with constant coefficients, i.e. $T = \{\mathfrak{z} = \sum_{l=1}^n b_l \partial_l \mid b_l \in \mathbb{C}\}$. For $I \in \mathcal{P}(\{1,\ldots,n\})$ we denote by $T_I \subset T$ the space of vector fields parallel to X_I which yields

$$T_I = \{\mathfrak{z} \in T \mid \mathrm{d}g(\mathfrak{z}) = 0 \text{ for all } g \in G_I\} = \Big\{ \mathfrak{z} = \sum_{\substack{l=1\\ l\notin I}}^{n} b_l \partial_l \mid b_l \in \mathbb{C} \Big\}.$$

- For $0 \le k \le n$ let $\overline{\Omega}^k$ denote the complex vector space of exterior k-forms on $\mathbb{C}^n$ with constant coefficients, i.e. $\overline{\Omega}^k = \Big\{ \sum_{\substack{\{l_1,\ldots,l_k\}\in\\ \mathcal{P}(\{1,\ldots,n\})}} c_{l_1\ldots l_k} \mathrm{d}z_{l_1} \wedge \ldots \wedge \mathrm{d}z_{l_k} \mid c_{l_1\ldots l_k} \in \mathbb{C} \Big\}$.

  We denote by ω_I the unique $(n-|I|)$-form in $\overline{\Omega}^{n-|I|}$ given by

$$\omega_I := 1 \cdot \mathrm{d}z_{l_1} \wedge \ldots \wedge \mathrm{d}z_{l_{n-|I|}} \quad \text{where } \{l_1,\ldots,l_{n-|I|}\} = \{1,\ldots,n\} \setminus I,\ l_1 < \ldots < l_{n-|I|}\,.$$

We set

$$\overline{\Omega}_I := \overline{\Omega}^{n-|I|} \Big/ \left\{ c \,\mathrm{d}g_1 \wedge \ldots \wedge \mathrm{d}g_{n-|I|} \;\middle|\; \begin{array}{l} c \in \mathbb{C},\, g_1, \ldots, g_{n-|I|} \in G \text{ and} \\ \exists\, k \in \{1, \ldots, n-|I|\} \text{ such that } g_k \in G_I \end{array} \right\}.$$

It is easy to see that

$$\overline{\Omega}_I \cong \{c \cdot \omega_I \mid c \in \mathbb{C}\}$$

which is consistent with our definition of $\overline{\Omega}_I$ in Definition 2.1.

- For $I \neq \{1, \ldots, n\}$ and $i \in \{1, \ldots, n\} \setminus I$ the following choice of a so called *edge framing* is made. This is a choice of a function and a vector field for every edge. We need

$$g_{I \cup \{i\}, I} \in G_{I \cup \{i\}} \setminus G_I \quad \text{and} \quad \mathfrak{z}_{I, I \cup \{i\}} \in T_I \setminus T_{I \cup \{i\}}\,.$$

So we choose

$$g_{I \cup \{i\}, I}(z_1, \ldots, z_n) = z_i \quad \text{and} \quad \mathfrak{z}_{I, I \cup \{i\}} = \partial_i\,.$$

Let

$$|i, I| := |\{k \in \{1, \ldots, n\} \setminus I \mid k < i\}|\,.$$

Then $g_{I \cup \{i\}, I}$ and $\mathfrak{z}_{I, I \cup \{i\}}$ fulfil

$$i_{\mathfrak{z}_{I, I \cup \{i\}}}((-1)^{|i,I|}\, \omega_I) = \omega_{I \cup \{i\}} \quad \text{and} \quad \mathrm{d}f_{I \cup \{i\}, I} \wedge ((-1)^{|i,I|}\, \omega_{I \cup \{i\}}) = \omega_I$$

where $i_{\mathfrak{z}}(\omega)$ is the contraction of the differential form ω with the vector field $\mathfrak{z}$.

- Using these identities of the edge framing, the $\mathbb{C}$-linear maps $\pi_{I, I \cup \{i\}} \colon \overline{\Omega}_{I \cup \{i\}} \to \overline{\Omega}_I$ and $\pi_{I \cup \{i\}, I} \colon \overline{\Omega}_I \to \overline{\Omega}_{I \cup \{i\}}$ are defined as follows:

$$\begin{aligned} \pi_{I, I \cup \{i\}}(\omega_{I \cup \{i\}}) &= (-1)^{|i,I|}\, \omega_I & &\text{fulfilling} & \pi_{I, I \cup \{i\}}(i_{\mathfrak{z}_{I, I \cup \{i\}}}(\omega_I)) &= \omega_I \\ \pi_{I \cup \{i\}, I}(\omega_I) &= (-1)^{|i,I|}\, \omega_{I \cup \{i\}} & &\text{fulfilling} & \mathrm{d}f_{I \cup \{i\}, I} \wedge \pi_{I \cup \{i\}, I}(\omega_I) &= \omega_I \end{aligned}$$

Given the quiver $\mathcal{U}_n$, we define the associated quiver $\mathscr{D}$-module $E\mathcal{U}_n$ as the quotient of

$$\bigoplus_{I \in \mathcal{P}(\{1, \ldots, n\})} (\mathscr{D} \otimes_{\mathbb{C}} \overline{\Omega}_I \otimes_{\mathbb{C}} V_I)$$

by the subsheaf whose sections over $U \subset \mathbb{C}^n$, open, are given by $\mathbb{C}$-linear combinations of the following elements

$$\text{(i)}\; P \cdot \mathfrak{z}_I \otimes \omega_I \otimes v_I - \sum_{\substack{i=1 \\ i \notin I}}^{n} \mathrm{d}g_{I \cup \{i\}, I}(\mathfrak{z}_I) \cdot P \otimes \pi_{I \cup \{i\}, I}(\omega_I) \otimes A_{I \cup \{i\}, I}(v_I)$$

$$\text{(ii)}\; P \cdot g_L \otimes \omega_L \otimes v_L - \sum_{\substack{l=1 \\ l \in L}}^{n} \mathrm{d}g_L(\mathfrak{z}_{L \setminus \{l\}, L}) \cdot P \otimes \pi_{L \setminus \{l\}, L}(\omega_L) \otimes A_{L \setminus \{l\}, L}(v_L)$$

where $I, L \in \mathcal{P}(\{1, \ldots, n\})$, $P \in \mathscr{D}(U)$, $\mathfrak{z}_I \in T_I$, $g_L \in G_L$, $v_I \in V_I$, $v_L \in V_L$. We note that for $I = \{1, \ldots, n\}$ and $L = \varnothing$ the elements are trivially 0.

Using the definitions of T_I, G_L, the edge framing and the π–maps, the sections of the subsheaf are generated by the following elements

$$\text{(i)}\ P\cdot\left(\sum_{\substack{i=1\\ i\notin I}}^{n} b_i\partial_i\right)\otimes\omega_I\otimes v_I-\sum_{\substack{i=1\\ i\notin I}}^{n} b_i\cdot P\otimes(-1)^{|i,I|}\,\omega_{I\cup\{i\}}\otimes A_{I\cup\{i\},I}(v_I)$$

$$\text{(ii)}\ P\cdot\left(\sum_{\substack{l=1\\ l\in L}}^{n} a_l z_l\right)\otimes\omega_L\otimes v_L-\sum_{\substack{l=1\\ l\in L}}^{n} a_l\cdot P\otimes(-1)^{|l,L\setminus\{l\}|}\,\omega_{L\setminus\{l\}}\otimes A_{L\setminus\{l\},L}(v_L)$$

where $I,L\in\mathcal{P}(\{1,\dots,n\})$, $P\in\mathscr{D}(U)$, $b_i,a_l\in\mathbb{C}$, $v_I\in V_I$, $v_L\in V_L$.

So, the sections of the subsheaf are generated by the following elements (set $I=L\setminus\{l\}$ in (ii))

$$\text{(i)}\ \ P\cdot\partial_i\otimes\omega_I\otimes v_I-P\otimes\omega_{I\cup\{i\}}\otimes(-1)^{|i,I|}\,A_{I\cup\{i\},I}(v_I)$$
$$\text{(ii)}\ P\cdot z_i\otimes\omega_{I\cup\{i\}}\otimes v_{I\cup\{i\}}-P\otimes\omega_I\otimes(-1)^{|i,I|}\,A_{I,I\cup\{i\}}(v_{I\cup\{i\}})$$

where $I\in\mathcal{P}(\{1,\dots,n\})\setminus\{1,\dots,n\}$, $i\in\{1,\dots,n\}\setminus I$, $P\in\mathscr{D}(U)$, $v_J\in V_J$.

Now, replace the quiver $(V_I,A_{I\cup\{i\},I},A_{I,I\cup\{i\}})$ by the quiver $(V_I,B_{I\cup\{i\},I},B_{I,I\cup\{i\}})$ where

$$B_{I\cup\{i\},I}:=(-1)^{|i,I|}\,A_{I\cup\{i\},I}\quad\text{and}\quad B_{I,I\cup\{i\}}:=(-1)^{|i,I|}\,A_{I,I\cup\{i\}}\,.$$

The quiver $(V_I,B_{I\cup\{i\},I},B_{I,I\cup\{i\}})$ fulfils the commutativity conditions, i. e. it is an object in $\mathcal{Q}ui_n$. Furthermore, it is immediately clear that we obtain the same quiver $\mathscr{D}$-module as the one above by Khoroshkin and Varchenko if we use the quiver $(V_I,B_{I\cup\{i\},I},B_{I,I\cup\{i\}})$ and our Definition 2.1.

Again, we note that Khoroshkin and Varchenko are using $\mathscr{D}$ as the sheaf of rings of linear algebraic differential operators over $\mathbb{C}^n$ instead of the one of analytic differential operators. As we are going to use the Riemann-Hilbert correspondence in the analytic setting in Chapter 3 and Chapter 4, we stated the Definition 2.1 of quiver $\mathscr{D}$-modules directly for $\mathscr{D}$ as sheaf of rings of analytic differential operators. However, we will come back to the algebraic vs. analytic topic, as the equality $\mathscr{D}^{\mathrm{an}}=\mathcal{O}^{\mathrm{an}}\otimes_{\mathcal{O}^{\mathrm{alg}}}\mathscr{D}^{\mathrm{alg}}$ obviously allows us to express every analytic quiver $\mathscr{D}$-module as the analytization of an algebraic quiver $\mathscr{D}$-module (see Corollary 4.10).

3. The 1-dimensional case

As a kind of warm up with quiver $\mathscr{D}$-modules we will restrict solely to dimension 1 in this chapter. First of all, let us observe how the quiver $\mathscr{D} = \mathscr{D}_{\mathbb{C}}$-module associated to an object from $\mathcal{Q}ui_1$ is given (cf. Definition 2.1):

Definition 3.1.
The functor E from the category of quiver representations $\mathcal{Q}ui_1$ into the category $\mathcal{M}od^S_{rh}(\mathscr{D})$ of regular singular holonomic $\mathscr{D}$-modules whose characteristic variety is contained in $\{(z,\xi) \mid z\xi = 0\}$ is given on objects as follows:
Let $\mathcal{V}_1 = \left(V_\varnothing \underset{A_{\varnothing,\alpha}}{\overset{A_{\alpha,\varnothing}}{\rightleftarrows}} V_\alpha \right)$ denote an object of the category $\mathcal{Q}ui_1$. We define the associated quiver $\mathscr{D}$-module $E\mathcal{V}_1$ as the quotient of

$$(\mathscr{D} \otimes_{\mathbb{C}} V_\varnothing) \oplus (\mathscr{D} \otimes_{\mathbb{C}} V_\alpha)$$

over the subsheaf $\mathcal{J}$. The sections of $\mathcal{J}$ over $U \subset \mathbb{C}^n$, open, are given by $\mathbb{C}$-linear combinations of the following elements

$$\begin{aligned} a\partial_z \otimes v_\varnothing - a \otimes A_{\alpha,\varnothing}(v_\varnothing) \quad &\text{and} \\ az \otimes v_\alpha - a \otimes A_{\varnothing,\alpha}(v_\alpha) \end{aligned}$$

where $a \in \mathscr{D}(U)$, $v_\varnothing \in V_\varnothing$ and $v_\alpha \in V_\alpha$.

From now on, we will only consider the situation locally at 0. In dimension 1 there exists a 1:1 correspondence between germs of holonomic $\mathscr{D}_{\mathbb{C}}$-modules and holonomic modules over the ring $\mathbb{C}\{z\}\langle\partial_z\rangle$. Therefore, we use $\mathscr{D} = \mathbb{C}\{z\}\langle\partial_z\rangle$ in the following and we consider the category $\mathcal{M}od_{\mathrm{rh}}(\mathscr{D})$ of regular singular holonomic $\mathscr{D}$-modules. The stratification in the index can be omitted in this local situation.

3.1. Quiver $\mathscr{D}$-modules, V-filtration and the Riemann-Hilbert correspondence

In this section, we are going to treat quiver $\mathscr{D}$-modules with a different approach which is only possible in dimension 1. Even though this does not work in higher dimension, it will provide an idea of how the general case in Chapter 4 works. The main ideas and results in this section are already known and may be found in [Bjö93, Section V.2] for instance. But partly only drafts of the proofs are given there.

Our approach here is with the help of V-filtration and the Riemann-Hilbert correspondence. In Chapter 4 we will use a theorem from [GGM85b], instead, which however involves (the general version of) the Riemann-Hilbert correspondence as well.

Now, let us recall some basic facts on the V-filtration:
The KASHIWARA-MALGRANGE V-FILTRATION (or short V-FILTRATION) $V_k\mathscr{D}$, $k \in \mathbb{Z}$, of $\mathscr{D}$ is defined by

$$P = \sum_{i=0}^{n} a_i(z)\partial_z^i \in V_k\mathscr{D} \;:\Longleftrightarrow\; \max_{i=0,\dots,n} \{i - \nu(a_i)\} \leq k$$

where $\nu(a_i)$ is the valuation of a_i. Then, similarly as for the standard filtration $F_\bullet\mathscr{D}$ of $\mathscr{D}$ (cf. proof of Theorem 2.4), one can define the notion of a good V-filtration $U_k\mathscr{M}$, $k \in \mathbb{Z}$, of a $\mathscr{D}$-module $\mathscr{M}$. In the following Proposition/Definition we state the main facts on the V-filtration that are important for us. This is based on [Sab93] and [MM04] where the proofs might be found.

Proposition/Definition 3.2.
Let $\mathscr{M}$ be a holonomic $\mathscr{D}$-module.

1. *Then $\mathscr{M}$ admits a good V-filtration.*
2. *Let $U_\bullet\mathscr{M}$ denote a good V-filtration of $\mathscr{M}$. A polynomial $b(s) \in \mathbb{C}[s]$ is called* BERNSTEIN-SATO POLYNOMIAL *for the good V-filtration $U_\bullet\mathscr{M}$ iff $b(s)$ is the unitary polynomial of smallest degree satisfying*

$$b(z\partial_z + k) \cdot U_k\mathscr{M} \subset U_{k-1}\mathscr{M} \quad \forall\, k \in \mathbb{Z}.$$

3. *Let $U_\bullet\mathscr{M}$ denote a good V-filtration of $\mathscr{M}$. Then $U_\bullet\mathscr{M}$ admits a Bernstein-Sato polynomial.*
4. *Let $U_\bullet\mathscr{M}$ denote a good V-filtration of $\mathscr{M}$. Then $\mathrm{gr}_k^U \mathscr{M} := U_k\mathscr{M}/U_{k-1}\mathscr{M}$ is a finitely generated $\mathbb{C}$-vector space for every $k \in \mathbb{Z}$.*
5. *There exists a unique good V-filtration of $\mathscr{M}$, denoted $V_\bullet^\Sigma\mathscr{M}$, such that the zeros of the corresponding Bernstein-Sato polynomial are contained in Σ (see Definition 1.3).*

The following corollary is an immediate consequence of Proposition/Definition 3.2 concerning the spectrum of $z\partial_z$ acting on $\mathrm{gr}_{-1}^{V^\Sigma}\mathscr{M}$.

Corollary 3.3.
Let $\mathscr{M}$ denote a holonomic $\mathscr{D}$-module and let $V_\bullet^\Sigma\mathscr{M}$ denote the unique good V-filtration of $\mathscr{M}$ from Proposition/Definition 3.2. Then the eigenvalues of $(z\partial_z)\cdot \_ : \mathrm{gr}_{-1}^{V^\Sigma}\mathscr{M} \to \mathrm{gr}_{-1}^{V^\Sigma}\mathscr{M}$ are contained in $\Sigma_1 = \Sigma + 1$.

Proof:
Let $b^\Sigma(s) \in \mathbb{C}[s]$ denote the Bernstein-Sato polynomial corresponding to the good V-filtration $V_\bullet^\Sigma\mathscr{M}$ of $\mathscr{M}$. By the definition of the Bernstein-Sato polynomial, the linear map

$$b^\Sigma(z\partial_z - 1)\cdot \_ : \mathrm{gr}_{-1}^{V^\Sigma}\mathscr{M} \to \mathrm{gr}_{-1}^{V^\Sigma}\mathscr{M}$$

is the zero map. Furthermore, the zeros of $b^\Sigma(s-1)$ are contained in $\Sigma+1 = \Sigma_1$. Let $m(s) \in \mathbb{C}[s]$ denote the minimal polynomial of the linear map

$$(z\partial_z)\cdot \_ : \mathrm{gr}_{-1}^{V^\Sigma}\mathscr{M} \to \mathrm{gr}_{-1}^{V^\Sigma}\mathscr{M}.$$

Then $m(s)$ divides $b^\Sigma(s-1)$. Consequently, the zeros of $m(s)$ and the eigenvalues of $(z\partial_z)\cdot \_$ acting on $\mathrm{gr}_{-1}^{V^\Sigma}\mathscr{M}$ are contained in Σ_1. This proves the claim. □

Treating regular singular holonomic $\mathscr{D}$-modules and the Riemann-Hilbert correspondence in dimension 1, one encounters usually the following two equivalent versions involving V-filtrations and quiver representations. Notes on the general version of the Riemann-Hilbert correspondence may be found after Theorem 4.2.

Theorem 3.4 ([Mal91], [Bjö93], [Dim04])
Let $\mathscr{M}$ denote a regular singular holonomic $\mathscr{D}$-module, i. e. an object in $\mathcal{M}od_{rh}(\mathscr{D})$. We define the linear maps

$$\begin{aligned} \mathrm{can}\colon \mathrm{gr}_{-1}^{V^\Sigma}\mathscr{M} &\to \mathrm{gr}_{0}^{V^\Sigma}\mathscr{M}, & \mathrm{var}\colon \mathrm{gr}_{0}^{V^\Sigma}\mathscr{M} &\to \mathrm{gr}_{-1}^{V^\Sigma}\mathscr{M} \quad \textit{and} \\ \mathrm{c}\colon \mathrm{gr}_{-1}^{V^\Sigma}\mathscr{M} &\to \mathrm{gr}_{0}^{V^\Sigma}\mathscr{M}, & \mathrm{v}\colon \mathrm{gr}_{0}^{V^\Sigma}\mathscr{M} &\to \mathrm{gr}_{-1}^{V^\Sigma}\mathscr{M} \end{aligned}$$

as follows:

$$\begin{aligned} \mathrm{can} &:= \partial_z \cdot \_\,, & \mathrm{var} &:= \sum_{k=1}^{\infty}\frac{(-2\pi i)^k}{k!}(z\partial_z)^{k-1} z\cdot \_ = z\cdot \sum_{k=1}^{\infty}\frac{(-2\pi i)^k}{k!}(\partial_z z)^{k-1}\cdot \_\,, \\ \mathrm{c} &:= \partial_z \cdot \_\,, & \mathrm{v} &:= z\cdot \_ \end{aligned}$$

(1) *The (covariant) functor $RH\colon \mathcal{M}od_{rh}(\mathscr{D}) \to \mathcal{C}_1$ from the category of regular singular holonomic $\mathscr{D}$-modules into the category $\mathcal{C}_1$ of quiver representations (see Definition 1.2) which associates to a regular singular holonomic $\mathscr{D}$-module $\mathscr{M}$ the object*

$$\mathrm{gr}_{-1}^{V^\Sigma}\mathscr{M} \underset{\mathrm{var}}{\overset{\mathrm{can}}{\rightleftarrows}} \mathrm{gr}_{0}^{V^\Sigma}\mathscr{M}$$

in $\mathcal{C}_1$ is an equivalence of categories.

(2) *The (covariant) functor $\widetilde{RH}\colon \mathcal{M}od_{rh}(\mathscr{D}) \to \mathcal{Q}ui_1^{\Sigma_1}$ from the category of regular singular holonomic $\mathscr{D}$-modules to the category $\mathcal{Q}ui_1^{\Sigma_1}$ of quiver representations (see Definition 1.3) which associates to a regular singular holonomic $\mathscr{D}$-module $\mathscr{M}$ the object*

$$\mathrm{gr}_{-1}^{V^\Sigma}\mathscr{M} \underset{\mathrm{v}}{\overset{\mathrm{c}}{\rightleftarrows}} \mathrm{gr}_{0}^{V^\Sigma}\mathscr{M}$$

in $\mathcal{Q}ui_1^{\Sigma_1}$ is an equivalence of categories.

First note, that the essential image of $\widetilde{RH}$ is indeed contained in $\mathcal{Q}ui_1^{\Sigma_1}$ by Corollary 3.3. Moreover, using the isomorphism of the categories $\mathcal{C}_1$ and $\mathcal{Q}ui_1^{\Sigma_1}$ from Theorem 1.11, we immediately see that statement (1) and (2) are equivalent as

$$RH = Q^- \circ \widetilde{RH} \quad \text{and} \quad \widetilde{RH} = \mathcal{G}^- \circ RH.$$

Proofs of statement (2) can be found in [Mal91, Section II.2] or [Bjö93, Section V.2]. Both very similar proofs directly show the equivalence of $\mathcal{M}od_{\mathrm{rh}}(\mathscr{D})$ with the quiver representation category $\mathcal{Q}ui_1^{\Sigma_1}$. And both do not use the equivalence with perverse sheaves as intermediate step. Statement (1) might be obtained by using an equivalence of the category of germs of perverse sheaves in $\mathbb{C}$ and the category $\mathcal{C}_1$ via vanishing and nearby cycles, and composing this functor with the de Rham functor (cf. [Dim04, Section 5.2 and Section 5.3]).

In the following we will apply the functor RH to quiver $\mathscr{D}$-modules. For this purpose we define a good V-filtration on quiver $\mathscr{D}$-modules in the next lemma and we examine gr_{-1}^V and gr_0^V. In Lemma 3.6 we will show that this V-filtration is already the unique good V-filtration $V_\bullet^{\Sigma_1}$ if the underlying quiver representation is an object in $\mathcal{Q}ui_1^{\Sigma_1}$.

Lemma 3.5.
Let $\mathcal{V}_1 = \left(V_\varnothing \underset{A_{\varnothing,\alpha}}{\overset{A_{\alpha,\varnothing}}{\rightleftarrows}} V_\alpha \right)$ *denote an object of the category* $\mathcal{Q}ui_1$ *and let*

$$\phi\colon (\mathscr{D} \otimes V_\varnothing) \oplus (\mathscr{D} \otimes V_\alpha) \twoheadrightarrow E\mathcal{V}_1$$

denote the projection of $(\mathscr{D} \otimes V_\varnothing) \oplus (\mathscr{D} \otimes V_\alpha)$ *onto* $E\mathcal{V}_1$*. We define the good V-filtration* $U_\bullet E\mathcal{V}_1$ *on* $E\mathcal{V}_1$ *as the filtration induced by* ϕ *via*

$$U_k E\mathcal{V}_1 := \phi\left((V_{k+1}\mathscr{D} \otimes V_\varnothing) \oplus (V_k\mathscr{D} \otimes V_\alpha) \right)$$

for $k \in \mathbb{Z}$*. Then we obtain*

$$\mathrm{gr}_{-1}^U E\mathcal{V} = V_\varnothing \quad \textit{and} \quad \mathrm{gr}_0^U E\mathcal{V} = V_\alpha .$$

Furthermore, the maps $\partial_z \cdot \_ \colon \mathrm{gr}_{-1}^U E\mathcal{V}_1 \to \mathrm{gr}_0^U E\mathcal{V}_1$ *and* $z \cdot \_ \colon \mathrm{gr}_0^U E\mathcal{V}_1 \to \mathrm{gr}_{-1}^U E\mathcal{V}_1$ *are given by* $A_{\alpha,\varnothing}$ *and* $A_{\varnothing,\alpha}$*, respectively.*

Proof:
$U_\bullet E\mathcal{V}_1$ is indeed a good V-filtration as it is induced by the exact sequence

$$(\mathscr{D} \otimes V_\varnothing) \oplus (\mathscr{D} \otimes V_\alpha) \xrightarrow{\phi} E\mathcal{V}_1 \to 0 .$$

The rest of the proof will be performed in three steps:

(i) First, we prove that $\mathrm{gr}_0^U E\mathcal{V}_1 = V_\alpha$: Let $P \in V_1\mathscr{D}$, $Q \in V_0\mathscr{D}$ and $v_\varnothing \in V_\varnothing$, $v_\alpha \in V_\alpha$. Thus

$$\phi((P \otimes v_\varnothing) \oplus (Q \otimes v_\alpha)) \in U_0 E\mathcal{V}_1 .$$

We denote by

$$\sigma_0\phi((P \otimes v_\varnothing) \oplus (Q \otimes v_\alpha)) =: \sigma_0\phi \begin{pmatrix} P \otimes v_\varnothing \\ Q \otimes v_\alpha \end{pmatrix}$$

its image in $\mathrm{gr}_0^U E\mathcal{V}_1$. Every element $P \in V_1\mathscr{D}$ may be written as $P = P_1\partial_z + P_2$ with $P_1, P_2 \in V_0\mathscr{D}$. Thus

$$\sigma_0\phi \begin{pmatrix} P \otimes v_\varnothing \\ Q \otimes v_\alpha \end{pmatrix} = \sigma_0\phi \begin{pmatrix} P_1\partial_z \otimes v_\varnothing \\ Q \otimes v_\alpha \end{pmatrix} + \underbrace{\sigma_0\,\phi \begin{pmatrix} P_2 \otimes v_\varnothing \\ 0 \end{pmatrix}}_{\in U_{-1}E\mathcal{V}_1} = \sigma_0\phi \begin{pmatrix} P_1\partial_z \otimes v_\varnothing \\ Q \otimes v_\alpha \end{pmatrix} =$$

$$= \sigma_0\phi \begin{pmatrix} 0 \\ P_1 \otimes A_{\alpha,\varnothing}(v_\varnothing) \end{pmatrix} + \sigma_0\phi \begin{pmatrix} 0 \\ Q \otimes v_\alpha \end{pmatrix} .$$

Let us focus on $\sigma_0\phi \begin{pmatrix} 0 \\ Q \otimes v_\alpha \end{pmatrix}$. The term $\sigma_0\phi \begin{pmatrix} 0 \\ P_1 \otimes A_{\alpha,\varnothing}(v_\varnothing) \end{pmatrix}$ may be treated similarly. Every element $Q \in V_0\mathscr{D}$ may be written as $Q = Q'z + p(\partial_z z)$ with $Q' \in V_0\mathscr{D}$, $p(s) \in \mathbb{C}[s]$.

Therefore, we obtain

$$\sigma_0\phi\begin{pmatrix}0\\ Q\otimes v_\alpha\end{pmatrix} = \underbrace{\sigma_0\,\phi\begin{pmatrix}0\\ Q'z\otimes v_\alpha\end{pmatrix}}_{\in U_{-1}E\mathcal{V}_1} + \sigma_0\phi\begin{pmatrix}0\\ p(\partial_z z)\otimes v_\alpha\end{pmatrix} = \sigma_0\phi\begin{pmatrix}0\\ p(\partial_z z)\otimes v_\alpha\end{pmatrix} =$$

$$= \sigma_0\phi\begin{pmatrix}0\\ 1\otimes p(A_{\alpha,\varnothing}A_{\varnothing,\alpha})(v_\alpha)\end{pmatrix}.$$

All in all, we have shown that every element $\sigma_0\phi\begin{pmatrix}P\otimes v_\varnothing\\ Q\otimes v_\alpha\end{pmatrix}$ of $\mathrm{gr}_0^U E\mathcal{V}_1$ can be expressed as an element of the form $\sigma_0\phi\begin{pmatrix}0\\ 1\otimes w_\alpha\end{pmatrix}$ where $w_\alpha \in V_\alpha$. This means $\mathrm{gr}_0^U E\mathcal{V}_1$ can be naturally identified with V_α as claimed.

(ii) Second, we prove that $\mathrm{gr}_{-1}^U E\mathcal{V}_1 = V_\varnothing$: Let $P\in V_0\mathscr{D}$, $Q\in V_{-1}\mathscr{D}$ and $v_\varnothing\in V_\varnothing$, $v_\alpha \in V_\alpha$. Thus

$$\phi((P\otimes v_\varnothing)\oplus(Q\otimes v_\alpha)) \in U_{-1}E\mathcal{V}_1\,.$$

We denote by

$$\sigma_{-1}\phi((P\otimes v_\varnothing)\oplus(Q\otimes v_\alpha)) =: \sigma_{-1}\phi\begin{pmatrix}P\otimes v_\varnothing\\ Q\otimes v_\alpha\end{pmatrix}$$

its image in $\mathrm{gr}_{-1}^U E\mathcal{V}_1$. Every element $Q\in V_{-1}\mathscr{D}$ may be written as $Q = Q'z$ with $Q'\in V_0\mathscr{D}$. Hence,

$$\sigma_{-1}\phi\begin{pmatrix}P\otimes v_\varnothing\\ Q\otimes v_\alpha\end{pmatrix} = \sigma_{-1}\phi\begin{pmatrix}P\otimes v_\varnothing\\ Q'z\otimes v_\alpha\end{pmatrix} = \sigma_{-1}\phi\begin{pmatrix}P\otimes v_\varnothing\\ 0\end{pmatrix} + \sigma_{-1}\phi\begin{pmatrix}Q'\otimes A_{\varnothing,\alpha}(v_\alpha)\\ 0\end{pmatrix}.$$

Let us focus on $\sigma_{-1}\phi\begin{pmatrix}P\otimes v_\varnothing\\ 0\end{pmatrix}$. The term $\sigma_{-1}\phi\begin{pmatrix}Q'\otimes A_{\varnothing,\alpha}(v_\alpha)\\ 0\end{pmatrix}$ may be treated similarly. Every element $P\in V_0\mathscr{D}$ might also be written as $P = P'z + q(z\partial_z)$ with $P'\in V_0\mathscr{D}$, $q(s)\in\mathbb{C}[s]$. Therefore,

$$\sigma_{-1}\phi\begin{pmatrix}P\otimes v_\varnothing\\ 0\end{pmatrix} = \underbrace{\sigma_{-1}\,\phi\begin{pmatrix}P'z\otimes v_\varnothing\\ 0\end{pmatrix}}_{\in U_{-2}E\mathcal{V}_1} + \sigma_{-1}\phi\begin{pmatrix}q(z\partial_z)\otimes v_\varnothing\\ 0\end{pmatrix} = \sigma_{-1}\phi\begin{pmatrix}q(z\partial_z)\otimes v_\varnothing\\ 0\end{pmatrix} =$$

$$= \sigma_{-1}\phi\begin{pmatrix}1\otimes q(A_{\varnothing,\alpha}A_{\alpha,\varnothing})(v_\varnothing)\\ 0\end{pmatrix}.$$

Altogether, we showed that every element $\sigma_{-1}\phi\begin{pmatrix}P\otimes v_\varnothing\\ Q\otimes v_\alpha\end{pmatrix}$ of $\mathrm{gr}_{-1}^U E\mathcal{V}_1$ can be expressed by an element of the form $\sigma_{-1}\phi\begin{pmatrix}1\otimes w_\varnothing\\ 0\end{pmatrix}$ for $w_\varnothing\in V_\varnothing$. Hence, $\mathrm{gr}_{-1}^U E\mathcal{V}_1$ can be naturally identified with $V_\varnothing$.

(iii) Given that $\mathrm{gr}_{-1}^U E\mathcal{V}_1 = V_\varnothing$ and $\mathrm{gr}_0^U E\mathcal{V}_1 = V_\alpha$, it is immediately clear from the definition of $E\mathcal{V}_1$ that $\partial_z\cdot\_ : \mathrm{gr}_{-1}^U E\mathcal{V}_1 \to \mathrm{gr}_0^U E\mathcal{V}_1$ and $z\cdot\_ : \mathrm{gr}_0^U E\mathcal{V}_1 \to \mathrm{gr}_{-1}^U E\mathcal{V}_1$ are given by $A_{\alpha,\varnothing}$ and $A_{\varnothing,\alpha}$, respectively. □

Lemma 3.6.
Let $\mathcal{V}_1 = \left(V_\varnothing \underset{A_{\varnothing,\alpha}}{\overset{A_{\alpha,\varnothing}}{\rightleftarrows}} V_\alpha \right)$ *denote an object of the category* $\mathcal{Q}ui_1^{\Sigma_1}$ *and let* $U_\bullet E\mathcal{V}_1$ *denote the good V-filtration of* $E\mathcal{V}_1$ *from Lemma 3.5. Then*

$$U_\bullet E\mathcal{V}_1 = V_\bullet^\Sigma E\mathcal{V}_1 \,,$$

i. e. $U_\bullet E\mathcal{V}_1$ *is already the unique good V-filtration from Proposition/Definition 3.2.*

Proof:
Let $m(s) \in \mathbb{C}[s]$ denote the minimal polynomial of $A_{\varnothing,\alpha}A_{\alpha,\varnothing}\colon V_\varnothing \to V_\varnothing$ and let $n(s) \in \mathbb{C}[s]$ denote the minimal polynomial of $A_{\alpha,\varnothing}A_{\varnothing,\alpha}\colon V_\alpha \to V_\alpha$. We set

$$b(s) := m(s+1) \cdot n(s+1) \,.$$

We claim that this unitary polynomial fulfils for all $k \in \mathbb{Z}$ that

$$b(z\partial_z + k) \cdot U_k E\mathcal{V}_1 \subset U_{k-1} E\mathcal{V}_1 \,,$$

and that the zeros of $b(s)$ are contained in Σ. This will be done in the following two steps:

1. The eigenvalues of $A_{\varnothing,\alpha}A_{\alpha,\varnothing}$ and $A_{\alpha,\varnothing}A_{\varnothing,\alpha}$ are contained in Σ_1 as $\mathcal{V}_1$ is an object in $\mathcal{Q}ui_1^{\Sigma_1}$. Therefore the zeros of $m(s+1)$, $n(s+1)$ and $b(s)$ are contained in Σ.
2. Fix $k \in \mathbb{Z}$. Let $P \in V_{k+1}\mathscr{D}$, $Q \in V_k\mathscr{D}$ and $v_\varnothing \in V_\varnothing$, $v_\alpha \in V_\alpha$. Then

$$\phi((P \otimes v_\varnothing) \oplus (Q \otimes v_\alpha)) =: \phi \begin{pmatrix} P \otimes v_\varnothing \\ Q \otimes v_\alpha \end{pmatrix} \in U_k E\mathcal{V}_1 .$$

The action of $b(z\partial_z + k)$ on $\phi((P \otimes v_\varnothing) \oplus (Q \otimes v_\alpha))$ is given by

$$b(z\partial_z + k) \cdot \phi \begin{pmatrix} P \otimes v_\varnothing \\ Q \otimes v_\alpha \end{pmatrix} = \phi \begin{pmatrix} n(z\partial_z + k + 1) m(z\partial_z + k + 1) P \otimes v_\varnothing \\ m(z\partial_z + k + 1) n(z\partial_z + k + 1) Q \otimes v_\alpha \end{pmatrix} .$$

We note that one finds $P' \in V_k\mathscr{D}$ and $Q' \in V_{k-1}\mathscr{D}$ such that

$$\begin{aligned} m(z\partial_z + k + 1) \cdot P &= P \cdot m(z\partial_z) + P' \quad \text{and} \\ n(z\partial_z + k + 1) \cdot Q &= Q \cdot n(z\partial_z + 1) + Q' = Q \cdot n(\partial_z z) + Q' . \end{aligned}$$

This yields

$$\begin{aligned} & b(z\partial_z + k) \cdot \phi \begin{pmatrix} P \otimes v_\varnothing \\ Q \otimes v_\alpha \end{pmatrix} = \\ &= \phi \begin{pmatrix} n(z\partial_z + k + 1) P m(z\partial_z) \otimes v_\varnothing \\ m(z\partial_z + k + 1) Q n(\partial_z z) \otimes v_\alpha \end{pmatrix} + \phi \begin{pmatrix} n(z\partial_z + k + 1) P' \otimes v_\varnothing \\ m(z\partial_z + k + 1) Q' \otimes v_\alpha \end{pmatrix} = \\ &= \underbrace{\phi \begin{pmatrix} n(z\partial_z + k + 1) P \otimes m(A_{\varnothing,\alpha}A_{\alpha,\varnothing})(v_\varnothing) \\ m(z\partial_z + k + 1) Q \otimes n(A_{\alpha,\varnothing}A_{\varnothing,\alpha})(v_\alpha) \end{pmatrix}}_{=0} + \phi \begin{pmatrix} n(z\partial_z + k + 1) P' \otimes v_\varnothing \\ m(z\partial_z + k + 1) Q' \otimes v_\alpha \end{pmatrix} = \\ &= \phi \begin{pmatrix} n(z\partial_z + k + 1) P' \otimes v_\varnothing \\ m(z\partial_z + k + 1) Q' \otimes v_\alpha \end{pmatrix} \in U_{k-1} E\mathcal{V}_1 . \end{aligned}$$

Hence, $b(z\partial_z + k) \cdot U_k E\mathcal{V}_1 \subset U_{k-1} E\mathcal{V}_1$.

In the case that $b(s)$ is already the Bernstein-Sato polynomial, i. e. it also fulfils the condition on the minimality of the degree, we are done. So assume that $b(s)$ is not the Bernstein-Sato polynomial. Then the proper Bernstein-Sato polynomial $b^{\Sigma}(s) \in \mathbb{C}[s]$ has smaller degree than $b(s)$. Using polynomial division we may write

$$b(s) = q(s) \cdot b^{\Sigma}(s) + r(s)$$

with $q(s), r(s) \in \mathbb{C}[s]$ and r has smaller degree than b^{Σ}. But $r = b - q \cdot b^{\Sigma}$ also fulfils the condition

$$r(z\partial_z + k) \cdot U_k E\mathcal{V}_1 \subset U_{k-1} E\mathcal{V}_1 \text{ for } k \in \mathbb{Z}$$

in contradiction to the minimality of the degree of b^{Σ}. Hence, $r \equiv 0$ and the zeros of $b^{\Sigma}(s)$ are contained in Σ, too. Hence, $U_{\bullet} E\mathcal{V}_1 = V_{\bullet}^{\Sigma} E\mathcal{V}_1$ in any case as claimed. ☐

Now, we can pursue with the 1-dimensional version of our Main Theorem 4.5. We show that every regular singular holonomic $\mathscr{D}$-module is isomorphic to a quiver $\mathscr{D}$-module, and in particular we obtain a quasi-inverse of the Riemann-Hilbert correspondence from Theorem 3.4.

Theorem 3.7.
Let $\mathcal{C}_1$ and $\mathcal{Q}ui_1^{\Sigma_1}$ denote the quiver representation categories from Chapter 1. Let $\mathcal{M}od_{rh}(\mathscr{D})$ denote the category of regular singular holonomic $\mathscr{D}$-modules. Then the diagram

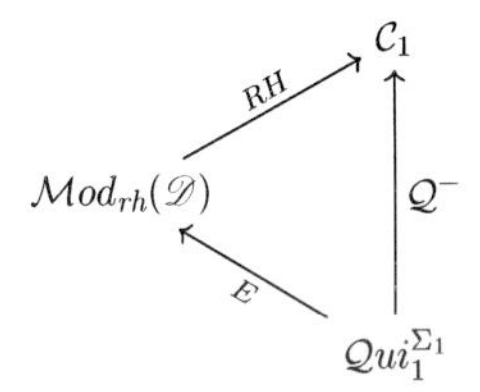

commutes. Furthermore, the functor $E\colon \mathcal{Q}ui_1^{\Sigma_1} \to \mathcal{M}od_{rh}(\mathscr{D})$ defines an equivalence of categories and E is quasi-inverse to $\widetilde{RH} = \mathcal{G}^- \circ RH$. In particular, E is essentially surjective. This means that the category of quiver $\mathscr{D}$-modules is exactly the category $\mathcal{M}od_{rh}(\mathscr{D})$, and every $\mathscr{D}$-module in $\mathcal{M}od_{rh}(\mathscr{D})$ is isomorphic to a quiver $\mathscr{D}$-module.

The definitions of the functors involved may be found in Theorem 1.11, Definition 3.1 and Theorem 3.4. We already noted before that $\widetilde{RH} = \mathcal{G}^- \circ RH$.

Proof:
On the level of morphisms the commutativity of the diagram is clear just by considering the definitions of the functors which are involved. On the level of objects the commutativity is easy to prove, too. We use the unique good V-filtration from Lemma 3.5 and Lemma 3.6, and we apply the Riemann-Hilbert correspondence. Then the definition of $\mathcal{Q}^-$ immediately shows us that $\mathcal{Q}^- = RH \circ E$. The commutativity of the diagram and the fact that $\mathcal{G}^-$ is the inverse of $\mathcal{Q}^-$, yields that the functor $\mathcal{G}^- \circ RH$ is a quasi-inverse of E. ☐

3.2. A complete list of examples

In Theorem 3.7 we have seen that $E\colon \mathcal{Qui}_1^{\Sigma_1} \to \mathcal{Mod}_{\mathrm{rh}}(\mathscr{D})$ is essentially surjective. We will use this statement to reprove in this section the following theorem which stems from Boutet de Monvel.

Theorem 3.8 ([Bou83])
Locally at 0 every regular singular holonomic $\mathscr{D}$-module is a direct sum of indecomposable submodules, each of them isomorphic to one of the following modules — $n \in \mathbb{N}^+$, $m \in \mathbb{N}_0$ and $\alpha \in \mathbb{C}\setminus\mathbb{Z}$:

$$^{\mathscr{D}}/_{\mathscr{D}(z\partial_z-\alpha)^n}\,,\quad ^{\mathscr{D}}/_{\mathscr{D}(z\partial_z)^n}\,,\quad ^{\mathscr{D}}/_{\mathscr{D}(\partial_z z)^n}\,,\quad ^{\mathscr{D}}/_{\mathscr{D}(z\partial_z)^m z}\,,\quad ^{\mathscr{D}}/_{\mathscr{D}(\partial_z z)^m \partial_z}$$

We note for $\alpha_1,\alpha_2 \in \mathbb{C}\setminus\mathbb{Z}$ that $^{\mathscr{D}}/_{\mathscr{D}(z\partial_z-\alpha_1)^n}$ and $^{\mathscr{D}}/_{\mathscr{D}(z\partial_z-\alpha_2)^n}$ are isomorphic if and only if $\alpha_1-\alpha_2 \in \mathbb{Z}$. Therefore, one might assume that $\alpha \in \Sigma_1\setminus\{0\}$ in Theorem 3.8.

Our approach to the proof of this theorem, is to use the equivalence E and to show that we can obtain all these five basic $\mathscr{D}$-modules as quiver $\mathscr{D}$-modules using indecomposable quiver representations. We note that this approach for the proof is similar to the one in [Bjö93, Section V.2]. But in [Bjö93] only a sketch is given and concrete computations of the quiver $\mathscr{D}$-modules are missing. So first of all, we need to analyse how our quiver representations can be decomposed:

Proposition 3.9.
Let $E \underset{v}{\overset{u}{\rightleftarrows}} F$ be an object in $\mathcal{Qui}_1$ ($\mathcal{Qui}_1^{\Sigma_1}$). Then $E \underset{v}{\overset{u}{\rightleftarrows}} F$ is isomorphic to a direct sum where each summand is one of the following indecomposable quiver representations — $n \in \mathbb{N}^+, m \in \mathbb{N}_0$:

$$\mathbb{C}^n \underset{\mathcal{S}_{\mu,n}}{\overset{\mathrm{Id}_n}{\rightleftarrows}} \mathbb{C}^n \cong \mathbb{C}^n \underset{\mathrm{Id}_n}{\overset{\mathcal{S}_{\mu,n}}{\rightleftarrows}} \mathbb{C}^n\,,\quad \mu \in \mathbb{C}\setminus\{0\}\ (\mu \in \Sigma_1\setminus\{0\})\,,$$

$$\mathbb{C}^n \underset{\mathcal{S}_{0,n}}{\overset{\mathrm{Id}_n}{\rightleftarrows}} \mathbb{C}^n\,,\quad \mathbb{C}^n \underset{\mathrm{Id}_n}{\overset{\mathcal{S}_{0,n}}{\rightleftarrows}} \mathbb{C}^n\,,\quad \mathbb{C}^{m+1} \underset{\binom{0_{1\times m}}{\mathrm{Id}_m}}{\overset{(\mathrm{Id}_m,0_{m\times 1})}{\rightleftarrows}} \mathbb{C}^m\,,\quad \mathbb{C}^m \underset{(\mathrm{Id}_m,0_{m\times 1})}{\overset{\binom{0_{1\times m}}{\mathrm{Id}_m}}{\rightleftarrows}} \mathbb{C}^{m+1}$$

where $\mathcal{S}_\mu = \mathcal{S}_{\mu,n}$ is the $n\times n$–matrix with $\mu \in \mathbb{C}$ on the main diagonal and 1 on the first upper off-diagonal, i. e.

$$\mathcal{S}_{\mu,n} = \begin{pmatrix} \lambda & 1 & & \\ & \ddots & \ddots & \\ & & \ddots & 1 \\ & & & \lambda \end{pmatrix}.$$

Proof:
First of all, we remark that all five quiver representations are indecomposable. The proof of the remaining assertions will be performed in several steps:

1. If $E=F=0$, we are done. So, assume that $E=0$ and $F\cong\mathbb{C}^n$ for $n\in\mathbb{N}^+$. Then:

$$E \underset{v}{\overset{u}{\rightleftarrows}} F \;\cong\; 0 \underset{0}{\overset{0}{\rightleftarrows}} \mathbb{C}^n \;\cong\; \bigoplus_{n\text{-times}} 0 \underset{0}{\overset{0}{\rightleftarrows}} \mathbb{C}$$

   In the same manner we obtain in the case $F=0$ and $E\cong\mathbb{C}^n$ for $n\in\mathbb{N}^+$ that

$$E \underset{v}{\overset{u}{\rightleftarrows}} F \;\cong\; \mathbb{C}^n \underset{0}{\overset{0}{\rightleftarrows}} 0 \;\cong\; \bigoplus_{n\text{-times}} \mathbb{C} \underset{0}{\overset{0}{\rightleftarrows}} 0\,.$$

In these three cases our claim is fulfilled. Therefore, we may assume that $\dim_{\mathbb{C}} E \geq 1$ and $\dim_{\mathbb{C}} F \geq 1$ in the following.

2. Let $a_1, \ldots, a_k$ denote the non-zero, pairwise disjoint eigenvalues of $v \circ u \colon E \to E$, and let E_{a_i} denote the generalized eigenspace of $v \circ u$ corresponding to a_i. Then E decomposes as

$$E = E_{a_1} \oplus \cdots \oplus E_{a_k} \oplus E_0 \,.$$

We use the convention that E_0 is the generalized eigenspace of $v \circ u$ corresponding to 0 if 0 is an eigenvalue of $v \circ u$, and $E_0 = 0$ otherwise. Using Proposition 1.4, we know that $a_1, \ldots, a_k$ are exactly the non-zero eigenvalues of $u \circ v \colon F \to F$. With the same notation as above we obtain a decomposition

$$F = F_{a_1} \oplus \cdots \oplus F_{a_k} \oplus F_0 \,.$$

We claim that we are given the following inclusions for $a \in \{a_1, \ldots, a_k, 0\}$:

$$u(E_a) \subseteq F_a \quad \text{and} \quad v(F_a) \subseteq E_a$$

Let $\alpha \in E_a$. Then,

$$0 = u \circ (v \circ u - a \cdot \mathrm{Id})^{\dim_{\mathbb{C}} E_a}(\alpha) = (u \circ v - a \cdot \mathrm{Id})^{\dim_{\mathbb{C}} E_a} u(\alpha)$$

which shows $u(\alpha) \in F_a$ and thus $u(E_a) \subseteq F_a$. The inclusion $v(F_a) \subseteq E_a$ is proved similarly. This yields that our quiver representation $E \underset{v}{\overset{u}{\rightleftarrows}} F$ decomposes as

$$E \underset{v}{\overset{u}{\rightleftarrows}} F \;=\; \bigoplus_{i=1}^{k} \left(E_{a_i} \underset{v_{a_i}}{\overset{u_{a_i}}{\rightleftarrows}} F_{a_i} \right) \oplus \; E_0 \underset{v_0}{\overset{u_0}{\rightleftarrows}} F_0$$

where we denote for $a \in \{a_1, \ldots, a_k, 0\}$ by u_a and v_a the restriction of u to E_a and v to F_a, respectively. So, we only need to consider a single summand of this direct sum in the following, i. e. a quiver representation $E \underset{v}{\overset{u}{\rightleftarrows}} F$ where $u \circ v$ and $v \circ u$ only admit one common eigenvalue $a \in \mathbb{C}$ ($a \in \Sigma_1$).

3. Let $E \underset{v}{\overset{u}{\rightleftarrows}} F$ be a quiver representation where $u \circ v$ and $v \circ u$ only admit one common eigenvalue $a \in \mathbb{C} \setminus \{0\}$ ($a \in \Sigma_1 \setminus \{0\}$). We assume that the Jordan form of $u \circ v$ has $l \in \mathbb{N}^+$ blocks $\mathcal{S}_{a,s_i}$ of size s_i for $i = 1, \ldots, l$. Let $e_{1,1}, \ldots, e_{1,s_1}, \ldots, e_{l,1}, \ldots, e_{l,s_l}$ denote the corresponding Jordan basis of E and let E_i denote the subspace of E generated by $e_{i,1}, \ldots, e_{i,s_i}$ for $i \in \{1, \ldots, l\}$. Clearly, u acts bijective as $a \neq 0$. So set

$$f_{i,j} := u(e_{i,j}) \quad \text{and} \quad F_i := u(E_i)$$

for $i = 1, \ldots, l$, $j = 1, \ldots, s_i$. This forms a basis of F. We see that our quiver representation w. r. t. these bases is given as

$$\bigoplus_{i=1}^{l} E_i \underset{v}{\overset{u}{\rightleftarrows}} \bigoplus_{i=1}^{l} F_i \;\cong\; \bigoplus_{i=1}^{l} \left(\mathbb{C}^{s_i} \underset{\mathcal{S}_{a,s_i}}{\overset{\mathrm{Id}_{s_i}}{\rightleftarrows}} \mathbb{C}^{s_i} \right).$$

So, in the case of one common eigenvalue $a \in \mathbb{C} \setminus \{0\}$ ($a \in \Sigma_1 \setminus \{0\}$) the claim is fulfilled.

4. We are left with the case when $E \underset{v}{\overset{u}{\rightleftarrows}} F$ is a quiver representation where $u \circ v$ and $v \circ u$ are nilpotent. Note that $u \circ v$ is nilpotent if and only if $v \circ u$ is nilpotent. Let

$$G := E \oplus F\,.$$

We define the map $s\colon G \to G$ by

$$\begin{aligned} s(e \oplus 0) &= 0 \oplus u(e) \quad \text{for } e \in E \text{ and} \\ s(0 \oplus f) &= v(f) \oplus 0 \quad \text{for } f \in F\,. \end{aligned}$$

Then, obviously the map s is nilpotent as well. To obtain a basis of G and E, F, we use the standard algorithm for nilpotent linear maps (see for example the proof of the Theorem "Normalform von nilpotenten Endomorphismen" in [Fis11, Subsection 4.2.6]). This algorithm yields the following:

We obtain a basis of G of the following form where $k \in \mathbb{N}^+$ is the degree of nilpotency of G:

$$\begin{array}{lll} w_1^{(k)}, s(w_1^{(k)}), \dots, s^{k-1}(w_1^{(k)}), & \dots, & w_{d_k}^{(k)}, s(w_{d_k}^{(k)}), \dots, s^{k-1}(w_{d_k}^{(k)}), \\ w_1^{(k-1)}, s(w_1^{(k-1)}), \dots, s^{k-2}(w_1^{(k-1)}), & \dots, & w_{d_{k-1}}^{(k-1)}, s(w_{d_{k-1}}^{(k-1)}), \dots, s^{k-2}(w_{d_{k-1}}^{(k-1)}), \\ \vdots & & \\ w_1^{(1)}, & \dots, & w_{d_1}^{(1)} \end{array}$$

Moreover, it holds that

$$G = \bigoplus_{i,j} Z_{ij} \quad \text{where} \quad Z_{ij} := \operatorname{span}\left(w_j^{(i)}, s(w_j^{(i)}), \dots, s^{i-1}(w_j^{(i)})\right)$$

for $i \in \{1, \dots, k\}$, $j \in \{1, \dots, d_i\}$. Furthermore, the vectors $w_j^{(i)}$ are elements of $\ker(s^i)$.

In our case the shape of s ensures that the vector $w_j^{(i)}$ can always be chosen such that it is of the form $e \oplus 0$ for some $e \in E$, or of the form $0 \oplus f$ for some $f \in F$. Therefore, all basis elements of G admit such a form. Hence, the direct sum decomposition of G induces naturally a direct sum decomposition of E and F

$$E = \bigoplus_{i,j} E_{ij} \quad \text{and} \quad F = \bigoplus_{i,j} F_{ij}$$

where E_{ij}, F_{ij} are the subspaces of E and F, respectively, induced by Z_{ij}. Possibly E_{1j} or F_{1j} is 0 which we permit in the following. Moreover, this decomposition of E and F is compatible with the maps u and v, i. e.

$$u(E_{ij}) \subseteq F_{ij} \quad \text{and} \quad v(F_{ij}) \subseteq E_{ij}\,.$$

This yields a decomposition of the quiver representation $E \underset{v}{\overset{u}{\rightleftarrows}} F$ into a direct sum:

$$E \underset{v}{\overset{u}{\rightleftarrows}} F \quad = \quad \bigoplus_{i,j} \left(E_{ij} \underset{v|_{F_{ij}}}{\overset{u|_{E_{ij}}}{\rightleftarrows}} F_{ij} \right)$$

So, in the following we consider a single summand $E_{ij} \underset{v|_{F_{ij}}}{\overset{u|_{E_{ij}}}{\rightleftarrows}} F_{ij}$.

First, let us treat the case $i = 1$, $j \in \{1, \ldots, d_1\}$ where two subcases may arise:

(i) The vector $w_j^{(1)}$ was chosen such that $w_j^{(1)} = e \oplus 0$ for some $e \in E$. Then the basis of E_{1j} is given by $\{e\}$ and $F_{1j} = 0$. In this case we have

$$E_{1j} \underset{v|_{F_{1j}}}{\overset{u|_{E_{1j}}}{\rightleftarrows}} F_{1j} \cong \mathbb{C} \underset{0}{\overset{0}{\rightleftarrows}} 0 .$$

(ii) The vector $w_j^{(1)}$ was chosen as $w_j^{(1)} = 0 \oplus f$ for some $f \in F$. Then the basis of F_{1j} is $\{f\}$ and $E_{1j} = 0$. We obtain in that case

$$E_{1j} \underset{v|_{F_{1j}}}{\overset{u|_{E_{1j}}}{\rightleftarrows}} F_{1j} \cong 0 \underset{0}{\overset{0}{\rightleftarrows}} \mathbb{C} .$$

Now, let $i \in \{2, \ldots, k\}$, $j \in \{1, \ldots, d_i\}$. The following four subcases may arise. We treat all of them exemplarily for $j = 1$.

(i) The vector $w_1^{(i)}$ was chosen as $w_1^{(i)} = e_1 \oplus 0$ for some $e_1 \in E$, and $s^{i-1}(w_1^{(i)}) = e \oplus 0$ for some $e \in E$. Then, E_{i1} admits the induced basis

$$\{e_1, (vu)(e_1), \ldots, (vu)^m(e_1)\}$$

where $m = \lceil \frac{i}{2} \rceil$, and F_{i1} has the basis

$$\{u(e_1), (uv)u(e_1), \ldots, (uv)^{m-1}u(e_1)\} .$$

W. r. t. these bases the quiver representation is given as

$$E_{i1} \underset{v|_{F_{i1}}}{\overset{u|_{E_{i1}}}{\rightleftarrows}} F_{i1} \cong \mathbb{C}^{m+1} \underset{\begin{pmatrix} 0_{1\times m} \\ \mathrm{Id}_m \end{pmatrix}}{\overset{(\mathrm{Id}_m, 0_{m\times 1})}{\rightleftarrows}} \mathbb{C}^m .$$

(ii) The vector $w_1^{(i)}$ was chosen as $w_1^{(i)} = 0 \oplus f_1$ for some $f_1 \in F$, and $s^{i-1}(w_1^{(i)}) = 0 \oplus f$ for some $f \in F$. In this case we obtain the following induced bases of F_{i1} and E_{i1}: F_{i1} has the basis

$$\{f_1, (uv)(f_1), \ldots, (uv)^m(f_1)\}$$

where $m = \lceil \frac{i}{2} \rceil$, and E_{i1} has the basis

$$\{v(f_1), (vu)v(f_1), \ldots, (vu)^{m-1}v(f_1)\} .$$

W. r. t. these bases the quiver representation is given as

$$E_{i1} \underset{v|_{F_{i1}}}{\overset{u|_{E_{i1}}}{\rightleftarrows}} F_{i1} \cong \mathbb{C}^{m} \underset{(\mathrm{Id}_m, 0_{m\times 1})}{\overset{\begin{pmatrix} 0_{1\times m} \\ \mathrm{Id}_m \end{pmatrix}}{\rightleftarrows}} \mathbb{C}^{m+1} .$$

(iii) The vector $w_1^{(i)}$ was chosen as $w_1^{(i)} = e_1 \oplus 0$ for some $e_1 \in E$, and $s^{i-1}(w_1^{(i)}) = 0 \oplus f$ for some $f \in F$. In this case E_{i1} has the induced basis

$$\{e_1, (vu)(e_1), \ldots, (vu)^n(e_1)\}$$

where $n = \frac{i}{2}$, and F_{i1} admits the basis

$$\{u(e_1), (uv)u(e_1), \ldots, (uv)^n u(e_1)\}.$$

W. r. t. these bases the quiver representation is given as

$$E_{i1} \underset{v|_{F_{i1}}}{\overset{u|_{E_{i1}}}{\rightleftarrows}} F_{i1} \cong \mathbb{C}^n \underset{S_{0,n}}{\overset{\mathrm{Id}_n}{\rightleftarrows}} \mathbb{C}^n .$$

(iv) The vector $w_1^{(i)}$ was chosen as $w_1^{(i)} = 0 \oplus f_1$ for some $f_1 \in F$, and $s^{i-1}(w_1^{(i)}) = e \oplus 0$ for some $e \in E$. Then, F_{i1} admits the induced basis

$$\{f_1, (uv)(f_1), \ldots, (uv)^n (f_1)\}$$

where $n = \frac{i}{2}$, and E_{i1} has the basis

$$\{v(f_1), (vu)v(f_1), \ldots, (vu)^n v(f_1)\}.$$

W. r. t. these bases the quiver representation is given as

$$E_{i1} \underset{v|_{F_{i1}}}{\overset{u|_{E_{i1}}}{\rightleftarrows}} F_{i1} \cong \mathbb{C}^n \underset{\mathrm{Id}_n}{\overset{S_{0,n}}{\rightleftarrows}} \mathbb{C}^n .$$

All in all, we obtain that in the case when $u \circ v$ and $v \circ u$ are nilpotent the claim is fulfilled as well. This completes the proof. □

The last step we have to take to reprove Theorem 3.8 is the computation of the quiver $\mathscr{D}$-modules that correspond to the list of five indecomposable quiver representations from Proposition 3.9 for $\mu \in \Sigma_1 \setminus \{0\}$. This is part of Example 3.2.1 to Example 3.2.5 where the computations will be performed in more generality for $\mu \neq 0$. The following table summarizes these examples and shows us that we obtain exactly the same list as Boutet de Monvel (Theorem 3.8).

Table 3.1.: **$\mathscr{D}$-module $E\mathcal{V}$ associated to the quiver representation $\mathcal{V}$**

Example	Quiver representation $\mathcal{V}$ ($n \in \mathbb{N}^+, m \in \mathbb{N}_0$)	Associated quiver $\mathscr{D}$-module $E\mathcal{V}$ is isomorphic to ...
3.2.1	$\mathbb{C}^n \underset{S_{\mu,n}}{\overset{\mathrm{Id}_n}{\rightleftarrows}} \mathbb{C}^n \cong \mathbb{C}^n \underset{\mathrm{Id}_n}{\overset{S_{\mu,n}}{\rightleftarrows}} \mathbb{C}^n$ for $\mu \neq 0$	$\mathscr{D}/\mathscr{D}(z\partial_z - \mu)^n \cong \mathscr{D}/\mathscr{D}(z\partial_z - (\mu-1))^n$
3.2.2	$\mathbb{C}^n \underset{S_{0,n}}{\overset{\mathrm{Id}_n}{\rightleftarrows}} \mathbb{C}^n$	$\mathscr{D}/\mathscr{D}(z\partial_z)^n$
3.2.3	$\mathbb{C}^n \underset{\mathrm{Id}_n}{\overset{S_{0,n}}{\rightleftarrows}} \mathbb{C}^n$	$\mathscr{D}/\mathscr{D}(\partial_z z)^n$
3.2.4	$\mathbb{C}^{m+1} \underset{\binom{0_{1\times m}}{\mathrm{Id}_m}}{\overset{(\mathrm{Id}_m, 0_{m\times 1})}{\rightleftarrows}} \mathbb{C}^m$	$\mathscr{D}/\mathscr{D}(\partial_z z)^m \partial_z$
3.2.5	$\mathbb{C}^m \underset{(\mathrm{Id}_m, 0_{m\times 1})}{\overset{\binom{0_{1\times m}}{\mathrm{Id}_m}}{\rightleftarrows}} \mathbb{C}^{m+1}$	$\mathscr{D}/\mathscr{D}(z\partial_z)^m z$

One also sees that non-isomorphic quiver representations may yield isomorphic $\mathscr{D}$-modules. For instance, the quiver representations

$$\mathbb{C}^n \underset{\mathcal{S}_1}{\overset{\mathrm{Id}}{\rightleftarrows}} \mathbb{C}^n \quad \text{and} \quad \mathbb{C}^n \underset{\mathcal{S}_2}{\overset{\mathrm{Id}}{\rightleftarrows}} \mathbb{C}^n$$

are non-isomorphic but the associated $\mathscr{D}$-modules

$$\mathscr{D}/\mathscr{D}(z\partial_z - 1)^n \quad \text{and} \quad \mathscr{D}/\mathscr{D}(z\partial_z - 2)^n$$

are isomorphic. This is the natural reason why the more restrictive category $\mathcal{Qui}_1^{\Sigma_1}$ appears in Theorem 3.7 instead of $\mathcal{Qui}_1$.

For the following examples let $V_\varnothing := \mathbb{C}^n$ and $V_\alpha := \mathbb{C}^m$ where $n, m \in \mathbb{N}_0$ and $\mathbb{C}^0 = 0$, i. e. we use the quiver representation

$$\mathcal{V} = \left(\mathbb{C}^n \underset{A_{\varnothing,\alpha}}{\overset{A_{\alpha,\varnothing}}{\rightleftarrows}} \mathbb{C}^m \right).$$

If $n = m = 0$, we immediately see that $E\mathcal{V} = 0$. In the other cases, we use the standard bases of $\mathbb{C}^n$ and $\mathbb{C}^m$ and we consider $A_{\alpha,\varnothing}$ and $A_{\varnothing,\alpha}$ as matrices of dimension $m \times n$ and $n \times m$, respectively. Moreover, we have a natural isomorphism from $(\mathscr{D} \otimes \mathbb{C}^n) \oplus (\mathscr{D} \otimes \mathbb{C}^m)$ to $\underbrace{\mathscr{D} \times \cdots \times \mathscr{D}}_{\text{(n+m)-times}} = \mathscr{D}^{n+m}$.

Using this notation, the $\mathscr{D}$-module $E\mathcal{V}$ is the cokernel of the $\mathscr{D}$-linear map

$$\begin{aligned} \varphi\colon \mathscr{D}^{n+m} &\longrightarrow \mathscr{D}^{n+m} \\ x &\longmapsto x^T \cdot \begin{pmatrix} \partial_z \cdot \mathrm{Id}_n & -A_{\alpha,\varnothing}^T \\ -A_{\varnothing,\alpha}^T & z \cdot \mathrm{Id}_m \end{pmatrix} \end{aligned}$$

where matrix multiplication is defined in the ordinary way.

3.2.1. Example: $\mathbb{C}^n \underset{\mathcal{S}_\mu}{\overset{\mathrm{Id}}{\rightleftarrows}} \mathbb{C}^n$ and $\mathbb{C}^n \underset{\mathrm{Id}}{\overset{\mathcal{S}_\mu}{\rightleftarrows}} \mathbb{C}^n$ for $\mu \neq 0$

Let $n \in \mathbb{N}^+$, $\mu \in \mathbb{C} \setminus \{0\}$ and set

$$\mathcal{V} := \left(\mathbb{C}^n \underset{\mathcal{S}_{\mu,n}}{\overset{\mathrm{Id}_n}{\rightleftarrows}} \mathbb{C}^n \right).$$

Note that $\mathcal{V} = \left(\mathbb{C}^n \underset{\mathcal{S}_\mu}{\overset{\mathrm{Id}}{\rightleftarrows}} \mathbb{C}^n \right)$ and $\widetilde{\mathcal{V}} := \left(\mathbb{C}^n \underset{\mathrm{Id}}{\overset{\mathcal{S}_\mu}{\rightleftarrows}} \mathbb{C}^n \right)$ are isomorphic via the isomorphism $(\mathrm{Id}, \mathcal{S}_\mu)$. Hence, $E\mathcal{V} \cong E\widetilde{\mathcal{V}}$ and we can solely focus on $\mathcal{V}$ for our computations.

Claim.

$$E\mathcal{V} \cong \mathscr{D}/\mathscr{D}(z\partial_z - \mu)^n$$

Proof:
Consider the diagram

$$\begin{array}{ccc} \mathscr{D} & \xrightarrow{\psi} & \mathscr{D} \\ \alpha\downarrow & & \downarrow\beta \\ \mathscr{D}^{2n} & \xrightarrow{\varphi} & \mathscr{D}^{2n} \end{array}$$

where $c := z\partial_z - \mu$ and

$$\begin{aligned} &\psi\colon \mathscr{D} \to \mathscr{D}, && \psi(t) = t \cdot c^n, \\ &\alpha\colon \mathscr{D} \to \mathscr{D}^{2n}, && \alpha(t) = t \cdot \left(z, cz, c^2 z, \dots, c^{n-1}z,\ 1, c, c^2, \dots, c^{n-1}\right)^T, \\ &\beta\colon \mathscr{D} \to \mathscr{D}^{2n}, && \beta(t) = (0, \dots, 0, \underbrace{t}_{\text{n-th entry}}, 0, \dots, 0)^T, \\ &\varphi\colon \mathscr{D}^{2n} \to \mathscr{D}^{2n}, && \varphi(x) = x^T \cdot \begin{pmatrix} \partial_z \cdot \mathrm{Id}_n & -\mathrm{Id}_n \\ -\mathcal{S}_{\mu,n}^T & z \cdot \mathrm{Id}_n \end{pmatrix}. \end{aligned}$$

One easily verifies that the diagram commutes. Therefore, the induced $\mathscr{D}$-linear morphism

$$\begin{aligned} \bar{\beta}\colon \mathscr{D}/\psi(\mathscr{D}) = \mathscr{D}/\mathscr{D}(z\partial_z - \mu)^n &\longrightarrow \mathscr{D}^{2n}/\varphi(\mathscr{D}^{2n}) = E\mathcal{V} \\ [P] &\longmapsto [(0, \dots, 0, \underbrace{P}_{\text{n-th entry}}, 0, \dots, 0)^T] \end{aligned}$$

is well-defined. Here, $[\,.\,]$ indicates the respective equivalence class. We show that $\bar{\beta}$ is actually an isomorphism which proves our claim.

(i) $\bar{\beta}$ is injective: Let $P \in \mathscr{D}$ with $\beta(P) = (0, \dots, 0, \underbrace{P}_{\text{n-th entry}}, 0, \dots, 0)^T \in \mathrm{im}(\varphi)$. Consider the equation

$$(0, \dots, 0, \underbrace{P}_{\text{n-th entry}}, 0, \dots, 0)^T = (\tilde{f}_1, \dots, \tilde{f}_n, f_1, \dots, f_n) \cdot \begin{pmatrix} \partial_z \cdot \mathrm{Id}_n & -\mathrm{Id}_n \\ -\mathcal{S}_{\mu,n}^T & z \cdot \mathrm{Id}_n \end{pmatrix}$$

where $\tilde{f}_i, f_i \in \mathscr{D}$ for $i \in \{1, \dots, n\}$. This equation yields $\tilde{f}_i = f_i \cdot z$ for $i = 1, \dots, n$. Hence,

$$f_n \cdot z\partial_z - f_n\mu = f_n(z\partial_z - \mu) = P \quad \text{and} \quad f_i \cdot z\partial_z - f_i\mu - f_{i+1} = f_i(z\partial_z - \mu) - f_{i+1} = 0$$

for $i = 1, \dots, n-1$. Step by step we obtain

$$P = f_n(z\partial_z - \mu) = f_{n-1}(z\partial_z - \mu)^2 = \ldots = f_1(z\partial_z - \mu)^n.$$

This means that $P \in \mathrm{im}(\psi)$ and therefore $\bar{\beta}$ is injective.

(ii) $\bar{\beta}$ is surjetive: Let $(P_1, \dots, P_n, Q_1, \dots, Q_n)^T \in \mathscr{D}^{2n}$. As

$$\varphi((Q_1, \dots, Q_n, 0, \dots, 0)^T) = (Q_1\partial_z, \dots, Q_n\partial_z, -Q_1, \dots, -Q_n)^T,$$

we obtain

$$[(P_1, \dots, P_n, Q_1, \dots, Q_n)^T] = [(P_1 + Q_1\partial_z, \dots, P_n + Q_n\partial_z, 0, \dots, 0)^T].$$

Furthermore, we have for $i \in \{2, \dots, n\}$ that

$$\varphi((0,\dots,0,\underbrace{z}_{\text{i-th entry}},0,\dots,0,\underbrace{1}_{\text{(n+i)-th entry}},0,\dots,0)^T) = (0,\dots,0,-1,\underbrace{z\partial_z - \mu}_{\text{i-th entry}},0,\dots,0)^T.$$

This implies inductively that

$$[(P_1 + Q_1\partial_z, \dots, P_n + Q_n\partial_z, 0, \dots, 0)^T] =$$
$$= \Big[(0,\dots,0,\underbrace{\sum_{i=1}^{n}(P_i + Q_i\partial_z)(z\partial_z - \mu)^{n-i}}_{\text{n-th entry}},0,\dots,0)^T\Big] = \bar{\beta}\left(\Big[\sum_{i=1}^{n}(P_i + Q_i\partial_z)(z\partial_z - \mu)^{n-i}\Big]\right)$$

and we see that $\bar{\beta}$ is surjective. □

We already stated that this also shows us that

$$E\widetilde{\mathcal{V}} \cong \mathscr{D}/\mathscr{D}(z\partial_z - \mu)^n .$$

We want to note here that one might also adapt the proof given in Example 3.2.3 to show that

$$E\widetilde{\mathcal{V}} \cong \mathscr{D}/\mathscr{D}(\partial_z z - \mu)^n .$$

This leads to the well-known isomorphism of analytic $\mathscr{D}$-modules for $\mu \neq 0$:

$$\mathscr{D}/\mathscr{D}(z\partial_z - \mu)^n \cong \mathscr{D}/\mathscr{D}(\partial_z z - \mu)^n = \mathscr{D}/\mathscr{D}(z\partial_z - (\mu - 1))^n$$

3.2.2. Example: $\mathbb{C}^n \underset{\mathcal{S}_0}{\overset{\mathrm{Id}}{\rightleftarrows}} \mathbb{C}^n$

Let $n \in \mathbb{N}^+$ and consider the quiver representation

$$\mathcal{V} := \left(\mathbb{C}^n \underset{\mathcal{S}_{0,n}}{\overset{\mathrm{Id}_n}{\rightleftarrows}} \mathbb{C}^n\right).$$

Examining Example 3.2.1 we see that all the calculations we made for $\mathbb{C}^n \underset{\mathcal{S}_\mu}{\overset{\mathrm{Id}}{\rightleftarrows}} \mathbb{C}^n$ with $\mu \neq 0$ are valid for $\mu = 0$ as well. This shows us that

$$E\mathcal{V} \cong \mathscr{D}/\mathscr{D}(z\partial_z)^n .$$

3.2.3. Example: $\mathbb{C}^n \underset{\mathrm{Id}}{\overset{\mathcal{S}_0}{\rightleftarrows}} \mathbb{C}^n$

Let $n \in \mathbb{N}^+$ and set

$$\mathcal{V} := \left(\mathbb{C}^n \underset{\mathrm{Id}_n}{\overset{\mathcal{S}_{0,n}}{\rightleftarrows}} \mathbb{C}^n\right).$$

Note that this quiver representation and the one from Example 3.2.2 are not isomorphic. However, we just need to refine our calculations from Example 3.2.1 a little bit to apply them here.

Claim.

$$\mathcal{EV} \cong \mathscr{D}/_{\mathscr{D}(\partial_z z)^n}$$

Proof:
To prove this we consider the diagram

$$\begin{array}{ccc} \mathscr{D} & \xrightarrow{\psi} & \mathscr{D} \\ {\scriptstyle\alpha}\downarrow & & \downarrow{\scriptstyle\beta} \\ \mathscr{D}^{2n} & \xrightarrow{\varphi} & \mathscr{D}^{2n} \end{array}$$

where $D := \partial_z z$ and

$$\begin{aligned}
&\psi\colon \mathscr{D} \to \mathscr{D}, && \psi(t) = t \cdot D^n\,,\\
&\alpha\colon \mathscr{D} \to \mathscr{D}^{2n}, && \alpha(t) = t \cdot \left(1, D, D^2, \ldots, D^{n-1},\ \partial_z, D\partial_z, D^2\partial_z, \ldots, D^{n-1}\partial_z\right)^T,\\
&\beta\colon \mathscr{D} \to \mathscr{D}^{2n}, && \beta(t) = (0, \ldots, 0, t)^T,\\
&\varphi\colon \mathscr{D}^{2n} \to \mathscr{D}^{2n}, && \varphi(x) = x^T \cdot \begin{pmatrix} \partial_z \cdot \mathrm{Id}_n & -\mathcal{S}_{0,n}^T \\ -\mathrm{Id}_n & z \cdot \mathrm{Id}_n \end{pmatrix}.
\end{aligned}$$

The diagram commutes and thus the induced $\mathscr{D}$-linear morphism

$$\begin{aligned}
\bar{\beta}\colon \mathscr{D}/_{\psi(\mathscr{D})} = \mathscr{D}/_{\mathscr{D}(\partial_z z)^n} &\longrightarrow \mathscr{D}^{2n}/_{\varphi(\mathscr{D}^{2n})} = \mathcal{EV}\\
[P] &\longmapsto [(0, \ldots, 0, P)^T]
\end{aligned}$$

is well-defined. $\bar{\beta}$ is even an isomorphism as we will show now:

(i) $\bar{\beta}$ is injective: Let $P \in \mathscr{D}$ with $(0, \ldots, 0, P)^T \in \operatorname{im}(\varphi)$. Then we have the identity

$$(0, \ldots, 0, P)^T = (f_1, \ldots, f_n, \tilde{f}_1, \ldots, \tilde{f}_n) \cdot \begin{pmatrix} \partial_z \cdot \mathrm{Id}_n & -\mathcal{S}_{0,n}^T \\ -\mathrm{Id}_n & z \cdot \mathrm{Id}_n \end{pmatrix}$$

where $f_i, \tilde{f}_i \in \mathscr{D}$ for $i \in \{1, \ldots, n\}$. We obtain $\tilde{f}_i = f_i \cdot \partial_z$ for $i = 1, \ldots, n$. Therefore,

$$f_n \cdot \partial_z z = P \quad \text{and} \quad -f_{i+1} + f_i \cdot \partial_z z = 0$$

for $i = 1, \ldots, n-1$. Inductively we obtain

$$P = f_n \cdot (\partial_z z) = f_{n-1} \cdot (\partial_z z)^2 = \ldots = f_1 (\partial_z z)^n\,.$$

Hence, $P \in \operatorname{im}(\psi)$ and $\bar{\beta}$ is injective.

(ii) $\bar{\beta}$ is surjective: Let $(P_1, \ldots, P_n, Q_1, \ldots, Q_n)^T \in \mathscr{D}^{2n}$. We use the identities

$$\begin{aligned}
&\varphi((0, \ldots, 0, P_1, \ldots, P_n)^T) = (-P_1, \ldots, -P_n, P_1 z, \ldots, P_n z)^T \quad \text{and}\\
&\varphi((0, \ldots, 0, \underbrace{1}_{\text{i-th entry}}, 0, \ldots, 0, \underbrace{\partial_z}_{\text{(n+i)-th entry}}, 0, \ldots, 0)^T) = (0, \ldots, 0, -1, \underbrace{\partial_z z}_{\text{(n+i)-th entry}}, 0, \ldots, 0)^T
\end{aligned}$$

for $i \in \{2, \ldots, n\}$.

This yields

$$[(P_1,\dots,P_n,Q_1,\dots,Q_n)^T] =$$
$$= \left[(0,\dots,0,\sum_{i=1}^{n}(Q_i+P_iz)(\partial_z z)^{n-i})^T\right] = \bar{\beta}\left(\left[\sum_{i=1}^{n}(Q_i+P_iz)(\partial_z z)^{n-i})^T\right]\right)$$

and we see that $\bar{\beta}$ is surjective. □

3.2.4. Example: $\mathbb{C}^{m+1} \underset{(0,\mathrm{Id}_m)^T}{\overset{(\mathrm{Id}_m,0)}{\rightleftarrows}} \mathbb{C}^m$

Let $m \in \mathbb{N}_0$ and consider the quiver representation

$$\mathcal{V} := \left(\mathbb{C}^{m+1} \underset{\begin{pmatrix} 0_{1\times m} \\ \mathrm{Id}_m \end{pmatrix}}{\overset{(\mathrm{Id}_m,0_{m\times 1})}{\rightleftarrows}} \mathbb{C}^m\right).$$

Claim.

$$E\mathcal{V} \cong \mathscr{D}/\mathscr{D}(\partial_z z)^m\partial_z$$

Proof:
For $m=0$ the $\mathscr{D}$-module $E\mathcal{V}$ corresponding to the quiver representation $\mathbb{C} \underset{0}{\overset{0}{\rightleftarrows}} 0$ is given by

$$E\mathcal{V} \cong \mathscr{D}/\mathscr{D}\partial_z$$

as asserted. So, in the following we focus on the case $m \in \mathbb{N}^+$. Therefor consider the diagram

$$\begin{array}{ccc} \mathscr{D} & \xrightarrow{\psi} & \mathscr{D} \\ \alpha\downarrow & & \downarrow\beta \\ \mathscr{D}^{2m+1} & \xrightarrow{\varphi} & \mathscr{D}^{2m+1} \end{array}$$

where $D := \partial_z z$ and

$$\begin{aligned}
&\psi\colon \mathscr{D} \to \mathscr{D}, && \psi(t) = t\cdot D^m\partial_z\,,\\
&\alpha\colon \mathscr{D} \to \mathscr{D}^{2m+1}, && \alpha(t) = t\cdot\left(D^m, D^{m-1},\dots,D,1,D^{m-1}\partial_z,D^{m-2}\partial_z,\dots,D\partial_z,\partial_z\right)^T,\\
&\beta\colon \mathscr{D} \to \mathscr{D}^{2m+1}, && \beta(t) = (t,0,\dots,0)^T,\\
&\varphi\colon \mathscr{D}^{2m+1} \to \mathscr{D}^{2m+1}, && \varphi(x) = x^T\cdot\begin{pmatrix} \partial_z\cdot\mathrm{Id}_{m+1} & \begin{matrix} -\mathrm{Id}_m \\ 0 \end{matrix} \\ 0,-\mathrm{Id}_m & z\cdot\mathrm{Id}_m \end{pmatrix}.
\end{aligned}$$

It is easily verified that the diagram commutes. The induced $\mathscr{D}$-linear morphism

$$\begin{aligned}
\bar{\beta}\colon \mathscr{D}/\psi(\mathscr{D}) = \mathscr{D}/\mathscr{D}(\partial_z z)^m\partial_z &\longrightarrow \mathscr{D}^{2m+1}/\varphi(\mathscr{D}^{2m+1}) = E\mathcal{V}\\
[P] &\longmapsto [(P,0,\dots,0)^T]
\end{aligned}$$

is therefore well-defined.

We are left with proving that $\bar{\beta}$ is an isomorphism:

(i) $\bar{\beta}$ is injective: Let $P \in \mathscr{D}$ with $(P, 0, \ldots, 0)^T \in \operatorname{im}(\varphi)$. We use the equation

$$(P, 0, \ldots, 0)^T = (f_1, \ldots, f_{m+1}, \tilde{f}_1, \ldots, \tilde{f}_m) \cdot \begin{pmatrix} \partial_z \cdot \mathrm{Id}_{m+1} & \begin{smallmatrix} -\mathrm{Id}_m \\ 0 \end{smallmatrix} \\ 0, -\mathrm{Id}_m & z \cdot \mathrm{Id}_m \end{pmatrix}$$

with $f_i, \tilde{f}_j \in \mathscr{D}$ for $i \in \{1, \ldots, m+1\}$, $j \in \{1, \ldots, m\}$. Hence,

$$P = f_1 \partial_z \quad \text{and} \quad f_{i+1} \cdot \partial_z = \tilde{f}_i$$

for $i = 1, \ldots, m$. This yields

$$-f_i + \tilde{f}_i z = -f_i + f_{i+1} \partial_z z = 0$$

for $i = 1, \ldots, m$. Step by step we obtain

$$P = f_1 \cdot \partial_z = f_2 \cdot (\partial_z z) \partial_z = \ldots = f_{m+1} \cdot (\partial_z z)^m \partial_z .$$

This means that $P \in \operatorname{im}(\psi)$ and $\bar{\beta}$ is injective.

(ii) $\bar{\beta}$ is surjective: Let $(P_1, \ldots, P_{m+1}, Q_1, \ldots, Q_m)^T \in \mathscr{D}^{2m+1}$. We use the identities

$$\varphi((Q_1, \ldots, Q_m, 0, \ldots, 0)^T) = (Q_1 \partial_z, \ldots, Q_m \partial_z, 0, -Q_1, \ldots, -Q_m)^T \quad \text{and}$$
$$\varphi((0, \ldots, 0, \underbrace{z}_{\text{i-th entry}}, 0, \ldots, 0, \underbrace{1}_{\text{(m+1+i)-th entry}}, 0, \ldots, 0)^T) = (0, \ldots, 0, \underbrace{z\partial_z}_{\text{i-th entry}}, -1, 0, \ldots, 0)^T$$

for $i \in \{1, \ldots, m\}$. This yields

$$[(P_1, \ldots, P_{m+1}, Q_1, \ldots, Q_m)^T] = \Big[(\sum_{i=1}^{m} (P_i + Q_i \partial_z)(z\partial_z)^{i-1} + P_{m+1}(z\partial_z)^m, 0, \ldots, 0)^T \Big] =$$
$$= \bar{\beta}\left(\Big[\sum_{i=1}^{m} (P_i + Q_i \partial_z)(z\partial_z)^{i-1} + P_{m+1}(z\partial_z)^m \Big] \right).$$

Thus, $\bar{\beta}$ is surjective. □

3.2.5. Example: $\mathbb{C}^m \underset{(\mathrm{Id}_m, 0)}{\overset{(0, \mathrm{Id}_m)^T}{\rightleftarrows}} \mathbb{C}^{m+1}$

Let $m \in \mathbb{N}_0$ and set

$$\mathcal{V} := \left(\mathbb{C}^m \underset{(\mathrm{Id}_m, 0_{m \times 1})}{\overset{\binom{0_{1 \times m}}{\mathrm{Id}_m}}{\rightleftarrows}} \mathbb{C}^{m+1} \right).$$

Claim.

$$E\mathcal{V} \cong \mathscr{D} / \mathscr{D}(z\partial_z)^m z$$

Proof:
For $m = 0$ we are given the quiver representation $0 \underset{0}{\overset{0}{\rightleftarrows}} \mathbb{C}$ and the corresponding $\mathscr{D}$-module $E\mathcal{V}$ is isomorphic to

$$E\mathcal{V} \cong \mathscr{D}/_{\mathscr{D}z}$$

as claimed. So let $m \in \mathbb{N}^+$ in the following. We consider the diagram

$$\begin{array}{ccc} \mathscr{D} & \xrightarrow{\psi} & \mathscr{D} \\ \downarrow{\scriptstyle\alpha} & & \downarrow{\scriptstyle\beta} \\ \mathscr{D}^{2m+1} & \xrightarrow{\varphi} & \mathscr{D}^{2m+1} \end{array}$$

where $D := z\partial_z$ and

$$\begin{aligned} &\psi\colon \mathscr{D} \to \mathscr{D}, && \psi(t) = t \cdot D^m z\,, \\ &\alpha\colon \mathscr{D} \to \mathscr{D}^{2m+1}, && \alpha(t) = t \cdot \left(D^{m-1}z, D^{m-2}z, \dots, Dz, z, D^m, D^{m-1}, \dots, D, 1\right)^T, \\ &\beta\colon \mathscr{D} \to \mathscr{D}^{2m+1}, && \beta(t) = (0, \dots, 0, \underbrace{t}_{\text{(m+1)-th entry}}, 0, \dots, 0)^T, \\ &\varphi\colon \mathscr{D}^{2m+1} \to \mathscr{D}^{2m+1}, && \varphi(x) = x^T \cdot \begin{pmatrix} \partial_z \cdot \mathrm{Id}_m & 0, -\mathrm{Id}_m \\ {-\mathrm{Id}_m \atop 0} & z \cdot \mathrm{Id}_{m+1} \end{pmatrix}. \end{aligned}$$

One easily verifies that the diagram commutes. Therefore, the induced morphism

$$\begin{aligned} \bar{\beta}\colon \mathscr{D}/_{\psi(\mathscr{D})} = \mathscr{D}/_{\mathscr{D}(z\partial_z)^m z} &\longrightarrow \mathscr{D}^{2m+1}/_{\varphi(\mathscr{D}^{2m+1})} = E\mathcal{V} \\ [P] &\longmapsto [(0, \dots, \underbrace{P}_{\text{(m+1)-th entry}}, 0, \dots, 0)^T] \end{aligned}$$

is well-defined. We will prove that $\bar{\beta}$ is even an isomorphism.

(i) $\bar{\beta}$ is injective: Let $P \in \mathscr{D}$ with $(0, \dots, 0, \underbrace{P}_{\text{(m+1)-th entry}}, 0, \dots, 0)^T \in \mathrm{im}(\varphi)$. Consider the equation

$$(0, \dots, 0, \underbrace{P}_{\text{(m+1)-th entry}}, 0, \dots, 0)^T = (\tilde{f}_1, \dots, \tilde{f}_m, f_1, \dots, f_{m+1}) \cdot \begin{pmatrix} \partial_z \cdot \mathrm{Id}_m & 0, -\mathrm{Id}_m \\ {-\mathrm{Id}_m \atop 0} & z \cdot \mathrm{Id}_{m+1} \end{pmatrix}$$

where $\tilde{f}_i, f_j \in \mathscr{D}$ for $i \in \{1, \dots, m\}$, $j \in \{1, \dots, m+1\}$. This yields

$$P = f_1 z \quad \text{and} \quad \tilde{f}_i = f_{i+1} z$$

for $i = 1, \dots, m$. Moreover, we have

$$\tilde{f}_i \partial_z - f_i = f_{i+1} z \partial_z - f_i = 0$$

for $i = 1, \dots, m$. Step by step we obtain

$$P = f_1 \cdot z = f_2 \cdot (z\partial_z) z = \ldots = f_{m+1} \cdot (z\partial_z)^m z\,.$$

Hence, $P \in \mathrm{im}(\psi)$ and therefore $\bar{\beta}$ is injective.

(ii) $\bar{\beta}$ is surjective: Let $(P_1, \ldots, P_m, Q_1, \ldots, Q_{m+1})^T \in \mathscr{D}^{2m+1}$. We use the identities

$$\varphi((0,\ldots,0,P_1,\ldots,P_m,0)^T) = (-P_1,\ldots,-P_m,P_1z,\ldots,P_mz,0)^T \quad \text{and}$$
$$\varphi((0,\ldots,0,\underbrace{1}_{\text{i-th entry}},0,\ldots,0,\underbrace{\partial_z}_{\text{(m+i)-th entry}},0,\ldots,0)^T) = (0,\ldots,0,\underbrace{\partial_z z}_{\text{(m+i)-th entry}},-1,0,\ldots,0)^T$$

for $i \in \{1,\ldots,m\}$. These yield inductively that

$$\begin{aligned}
&[(P_1,\ldots,P_m,Q_1,\ldots,Q_{m+1})^T] = \\
&= \Big[(0,\ldots,0,\underbrace{\sum_{i=1}^{m}(Q_i+P_iz)(\partial_z z)^{i-1}+Q_{m+1}(\partial_z z)^m}_{\text{(m+1)-th entry}},0,\ldots,0)^T\Big] = \\
&= \bar{\beta}\left(\left[\sum_{i=1}^{m}(Q_i+P_iz)(\partial_z z)^{i-1}+Q_{m+1}(\partial_z z)^m\right]\right).
\end{aligned}$$

We see that $\bar{\beta}$ is surjective. □

4. The general case

This is the main chapter of the present work. In Section 4.2 we show that in fact every regular singular holonomic $\mathscr{D}$-module whose characteristic variety is contained in Δ_S is isomorphic to a quiver $\mathscr{D}$-module. Moreover, we gain an explicit and simple quasi-inverse of the functor $\mathscr{A}$. The functor $\mathscr{A}$ yields an equivalence between $\mathcal{M}od^S_{\mathrm{rh}}(\mathscr{D})$ and $\mathcal{C}_n$ (see Theorem 4.2). Finally, in Corollary 4.10, we conclude that every object in $\mathcal{M}od^S_{\mathrm{rh}}(\mathscr{D})$ is isomorphic to the analytization of a regular singular holonomic algebraic $\mathscr{D}$-module.

First, we state the equivalence between $\mathcal{M}od^S_{\mathrm{rh}}(\mathscr{D})$ and $\mathcal{C}_n$ which is an important piece for Theorem 4.5. As the Riemann-Hilbert correspondence forms a part of this equivalence, we give some general remarks on it afterwards. In Section 4.1 we will prove some statements which are helpful for the proof of Theorem 4.5, whereas Section 4.2 is devoted to this proof.

Let us clarify some notational facts: We continue to use the notation stated at the beginning of Chapter 2. Furthermore, let $\iota\colon U \hookrightarrow X$ denote the inclusion for an open subset U of $\mathbb{C}^n$. Then Γ_U is the functor which maps sheaves on $\mathbb{C}^n$ to sheaves on $\mathbb{C}^n$ defined by

$$\Gamma_U := \iota_* \iota^{-1}.$$

Moreover, let

$$\mathbb{C}^n = \prod_{i=1}^{n} \mathbb{C}_i \quad \text{and} \quad W_i := \mathbb{C}_i \setminus \mathbb{R}_0^+$$

here. And for $I \in \mathcal{P}(\{1,\ldots,n\})$ set

$$\Lambda_I := \sum_{k\in I} \Gamma_{\mathbb{C}_k \times \prod_{\substack{i=1\\ i\neq k}}^{n} W_i} \mathcal{O} \quad \text{and} \quad \mathcal{O}_I := \frac{\Gamma_{\prod_{i=1}^{n} W_i} \mathcal{O}}{\Lambda_I}.$$

Example 4.1.
For $n = 1$,

$$\mathcal{O}_\varnothing = \Gamma_{\mathbb{C}\setminus\mathbb{R}_0^+} \mathcal{O} \quad \text{and} \quad \mathcal{O}_{\{1\}} = \frac{\Gamma_{\mathbb{C}\setminus\mathbb{R}_0^+}\mathcal{O}}{\mathcal{O}}.$$

For $n = 2$,

$$\mathcal{O}_\varnothing = \Gamma_{W_1\times W_2}\mathcal{O}, \quad \mathcal{O}_{\{1,2\}} = \frac{\Gamma_{W_1\times W_2}\mathcal{O}}{\Gamma_{\mathbb{C}_1\times W_2}\mathcal{O} + \Gamma_{W_1\times \mathbb{C}_2}\mathcal{O}},$$

$$\mathcal{O}_{\{1\}} = \frac{\Gamma_{W_1\times W_2}\mathcal{O}}{\Gamma_{\mathbb{C}_1\times W_2}\mathcal{O}} \quad \text{and} \quad \mathcal{O}_{\{2\}} = \frac{\Gamma_{W_1\times W_2}\mathcal{O}}{\Gamma_{W_1\times \mathbb{C}_2}\mathcal{O}}.$$

The following theorem of A. Galligo, M. Granger and Ph. Maisonobe will be important for our computations:

Theorem 4.2 ([GGM85a], [GGM85b])
The contravariant functor $\mathscr{A}$ from $\mathcal{M}od^S_{rh}(\mathscr{D})$ to $\mathcal{C}_n$

$$\mathscr{A}\colon \mathcal{M}od^S_{rh}(\mathscr{D}) \longrightarrow \mathcal{C}_n$$
$$\mathscr{M} \longmapsto \mathrm{Hom}_{\mathscr{D}_{X,0}}(\mathscr{M}_0, \mathcal{O}_{I,0}) \underset{\mathrm{var}_{I,i}}{\overset{\mathrm{can}_{I,i}}{\rightleftarrows}} \mathrm{Hom}_{\mathscr{D}_{X,0}}(\mathscr{M}_0, \mathcal{O}_{I\cup\{i\},0})$$

establishes an equivalence of categories. $\mathrm{can}_{I,i}$ *is the canonical map or quotient map which sends solutions with values in* $\mathcal{O}_{I,0}$ *to solutions with values in* $\mathcal{O}_{I\cup\{i\},0}$. $\mathrm{var}_{I,i}$ *is the variation around* $z_i = 0$, *i. e. we have*

$$\mathrm{var}_{I,i}(F) = M_i F - F \quad \text{for} \quad F \in \mathrm{Hom}_{\mathscr{D}_{X,0}}(\mathscr{M}_0, \mathcal{O}_{I\cup\{i\},0})$$

where $M_i F$ *is the class of a representative of* F *after analytic continuation around the axis* $z_i = 0$. *A* $\mathscr{D}$*-linear morphism* $\phi\colon \mathscr{M} \to \mathscr{N}$ *in* $\mathcal{M}od^S_{rh}(\mathscr{D})$ *is mapped under* $\mathscr{A}$ *to the morphism*

$$\big(\mathrm{Hom}_{\mathscr{D}_{X,0}}(\phi_0, \mathcal{O}_{I,0})\big) \quad \text{in} \quad \mathcal{C}_n$$

where $\mathrm{Hom}_{\mathscr{D}_{X,0}}(\phi_0, \mathcal{O}_{I,0})\colon \mathrm{Hom}_{\mathscr{D}_{X,0}}(\mathscr{N}_0, \mathcal{O}_{I,0}) \to \mathrm{Hom}_{\mathscr{D}_{X,0}}(\mathscr{M}_0, \mathcal{O}_{I,0})$ *is given by* $g \mapsto g \circ \phi_0$.

Parts of the proof are summarized in the following. But let us first remark some general facts on the Riemann-Hilbert correspondence and its relation to the previous theorem:

The Riemann-Hilbert correspondence as it was proven independently in 1980 by M. Kashiwara (see [Kas84]) and Z. Mebkhout (see [Meb84]) gives for a complex manifold X an equivalence between the derived category $\mathrm{D}^b_{\mathrm{rh}}(\mathscr{D}_X)$ of left $\mathscr{D}_X$-modules consisting of bounded complexes whose cohomology groups are regular holonomic, and the derived category $\mathrm{D}^b_{\mathrm{c}}(\mathbb{C}_X)$ of sheaves of $\mathbb{C}$-vector spaces on X consisting of bounded complexes whose cohomology groups are constructible. This equivalence might be proved using the covariant de Rham functor

$$\mathrm{DR}(\mathscr{M}^\bullet) = \mathbf{R}\,\mathrm{Hom}_{\mathscr{D}_X}(\mathcal{O}_X, \mathscr{M}^\bullet)$$

or the contravariant solutions functor

$$\mathrm{Sol}(\mathscr{M}^\bullet) = \mathbf{R}\,\mathrm{Hom}_{\mathscr{D}_X}(\mathscr{M}^\bullet, \mathcal{O}_X)$$

which fulfil

$$\mathbf{D}\mathrm{Sol}(\mathscr{M}^\bullet) = \mathrm{Sol}(\mathbb{D}\mathscr{M}^\bullet) \cong \mathrm{DR}(\mathscr{M}^\bullet)\,.$$

$\mathbf{D}$ is the duality functor in the derived category $\mathrm{D}^b_{\mathrm{c}}(\mathbb{C}_X)$ and $\mathbb{D}$ is the duality functor in the derived category $\mathrm{D}^b_{\mathrm{rh}}(\mathscr{D}_X)$.

Taking complexes concentrated in degree 0, the Riemann-Hilbert correspondence yields an equivalence between the category $\mathcal{M}od_{\mathrm{rh}}(\mathscr{D}_X)$ of regular holonomic $\mathscr{D}_X$-modules and the category $\mathrm{Perv}(X)$ of perverse sheaves on X. A proof hereof might be found in [HTT08, Section 7.2] or [Meb89, Section II.7]. Let $\mathscr{M}$ be an object in $\mathcal{M}od_{\mathrm{rh}}(\mathscr{D}_X)$ and let a Whitney stratification $\bigcup_{j\in J} X_j = X$ of X be given such that

$$\mathrm{Char}(\mathscr{M}) \subset \bigcup_{j\in J} T^*_{X_j} X\,.$$

Then, this is also the stratification used for the support and cosupport condition of the corresponding perverse sheaf under the Riemann-Hilbert correspondence. So, the Riemann-Hilbert correspondence "transports" this stratification from one side to another (cf. proof in [Meb89]). For this reason the stratification is an important part of the definition of $\mathcal{M}od^S_{\mathrm{rh}}(\mathscr{D})$ as it determines the essential image under the Riemann-Hilbert correspondence.

In the 1-dimensional case (see Chapter 3) we have already seen that besides this equivalence of the category of regular holonomic $\mathscr{D}_X$-modules and the one of perverse sheaves on X, other equivalences of regular holonomic $\mathscr{D}_X$-modules with categories of quiver representations exist. These are sometimes called Riemann-Hilbert correspondences as well. As noted there in dimension 1 some of these equivalences are proven without the use of perverse sheaves, whereas others of these Riemann-Hilbert correspondences use the "original" Riemann-Hilbert correspondence as stated above and compose it with an equivalence of perverse sheaves with quiver representations. This is the approach of Galligo, Granger and Maisonobe in [GGM85a]:

In their paper [GGM85a], Galligo, Granger and Maisonobe prove that the category of perverse sheaves in $\mathbb{C}^n$ with respect to the normal crossing stratification Ξ is equivalent to the category of quiver representations $\mathcal{C}_n$. They establish a functor

$$\alpha\colon \mathrm{Perv}^{\Xi}(\mathbb{C}^n) \to \mathcal{C}_n$$

which generalises the following statement of B. Malgrange in dimension 1: A $\mathscr{D}$-module on a small disc with regular singularity in 0 is loosely speaking completely determined by the correlation between its holomorphic solutions on $\mathbb{C}\backslash\mathbb{R}_0^+$ and which solutions can or cannot be holomorphically extended to $\mathbb{R}_0^+$. More precisely given a disc $D \subset \mathbb{C}$ the finite dimensional vector spaces

$$\mathrm{Hom}_{\mathscr{D}}(\mathscr{M}(D), \mathcal{O}(D \setminus \mathbb{R}_0^+)) \quad \text{and} \quad \mathrm{Hom}_{\mathscr{D}}\left(\mathscr{M}(D), \frac{\mathcal{O}(D \setminus \mathbb{R}_0^+)}{\mathcal{O}(D)}\right)$$

together with the projection and variation morphism completely determine the $\mathscr{D}$-module.
For a given object $\mathcal{F}^\bullet$ in $\mathrm{Perv}^{\Xi}(\mathbb{C}^n)$, Galligo, Granger and Maisonobe consider $\mathbf{R}\Gamma_Z\mathcal{F}^\bullet$ where $Z \subset \mathbb{C}^n$ is the product of n spaces of the form $\mathbb{C}$, $\mathbb{C}\backslash\mathbb{R}_0^+$ or $\mathbb{R}_0^+$. The functor Γ_Z was defined above for Z open. For Z closed it is the functor of sections supported by Z, and via $\Gamma_{Z_1}\Gamma_{Z_2} = \Gamma_{Z_1 \cap Z_2}$ it is defined for any locally closed set Z. Using distinguished triangles involving $\mathbf{R}\Gamma_Z\mathcal{F}^\bullet$'s they are able to show that $\mathbf{R}\Gamma_Z\mathcal{F}^\bullet$ is concentrated in one degree if Z is the product of n spaces of the form $\mathbb{C} \setminus \mathbb{R}_0^+$ or $\mathbb{R}_0^+$. Moreover, in this case $\mathbf{R}\Gamma_Z\mathcal{F}^\bullet$ is isomorphic to a sheaf which is locally constant on certain strata and $\mathbf{R}\Gamma_Z\mathcal{F}^\bullet$ can be completely recovered from its restriction to these strata. So, using the fibre of $\mathbf{R}\Gamma_Z\mathcal{F}^\bullet$ at 0 which is a $\mathbb{C}$-vector space and "transition maps" between the strata, it is possible to encode $\mathcal{F}^\bullet$ as an object of $\mathcal{C}_n$ which defines the functor α. They prove the equivalence of categories by description of an explicit quasi-inverse β.

Now, using the composition of the functor α and Sol one obtains an equivalence of the categories $\mathcal{M}od^S_{\mathrm{rh}}(\mathscr{D})$ and $\mathcal{C}_n$. Namely the functor $\mathscr{A}$ given in Theorem 4.2 is precisely naturally isomorphic to $\alpha \circ \mathrm{Sol}$. This is proved in [GGM85b]. It can be seen by observing that the functors $\mathbf{R}\Gamma_Z$ and $\mathbf{R}\,\mathrm{Hom}_{\mathscr{D}_X}(\mathscr{M}, \mathcal{O}_X)$ commute. Using [GGM85a] ($\mathbf{R}\,\mathrm{Hom}_{\mathscr{D}_X}(\mathscr{M}, \mathcal{O}_X)$ is a perverse sheaf) and the fact that $\mathbf{R}\Gamma_Z\mathcal{O}_X$ is concentrated in one degree, one may drop the $\mathbf{R}$'s and obtains the functor $\mathscr{A}$.

4.1. Properties of $\mathcal{O}_{I,0}$

Before applying the functor $\mathscr{A}$ to our quiver $\mathscr{D}$-modules we state some properties of $\mathcal{O}_{I,0}$ to simplify the arguments later in Section 4.2.

Obviously, $(\Gamma_{\prod_{i=1}^n W_i}\mathcal{O})_0$ and $\Lambda_{I,0}$ are unitary, commutative rings w. r. t. addition and multiplication of functions. And $\mathcal{O}_{I,0}$ is an abelian group w. r. t. addition of functions and a unitary left $\mathscr{D}_0$-module. Nevertheless, in general the multiplication of functions in $\mathcal{O}_{I,0}$ is not well-defined. Thus, $\mathcal{O}_{I,0}$ is not a ring.

First, consider the action of $\mathscr{D}_0$ on $\mathcal{O}_{I,0}$:

Lemma 4.3.
For $I \in \mathcal{P}(\{1,\dots,n\})$ let $\mathcal{O}_I$ as above. Then,

(i) z_j acts bijective on $\mathcal{O}_{I,0}$ if and only if $j \notin I$.

(ii) ∂_j acts bijective on $\mathcal{O}_{I,0}$ if and only if $j \in I$.

Proof:
Let $j \in \mathcal{P}(\{1,\dots,n\})$.

(i) For $k = 1,\dots,n$ we use Z_k as dummy for $\mathbb{C}_k$ or W_k. Consider the following diagram:

$$\begin{array}{ccccccccc}
0 & \longrightarrow & \Lambda_{I,0} & \hookrightarrow & (\Gamma_{\prod_{i=1}^n W_i}\mathcal{O})_0 & \twoheadrightarrow & \mathcal{O}_{I,0} & \longrightarrow & 0 \\
 & & \downarrow z_j\cdot & \circlearrowleft & \downarrow z_j\cdot & & \downarrow z_j\cdot & & \\
0 & \longrightarrow & \Lambda_{I,0} & \hookrightarrow & (\Gamma_{\prod_{i=1}^n W_i}\mathcal{O})_0 & \twoheadrightarrow & \mathcal{O}_{I,0} & \longrightarrow & 0
\end{array}$$

The rows are exact sequences and the square clearly commutes as indicated. This defines the action of z_j on $\mathcal{O}_{I,0} \cong \frac{(\Gamma_{\prod_{i=1}^n W_i}\mathcal{O})_0}{\Lambda_{I,0}}$. The inverse $\frac{1}{z_j}$ of z_j fulfils

$$\frac{1}{z_j} \in (\Gamma_{\prod_{i=1}^n Z_i}\mathcal{O})_0 \text{ if and only if } Z_j = W_j\,.$$

This means z_j acts bijective on $(\Gamma_{\prod_{i=1}^n Z_i}\mathcal{O})_0$ if and only if $Z_j = W_j$. Recalling that $\Lambda_{I,0} = (\sum_{k\in I}\Gamma_{\mathbb{C}_k\times\prod_{i=1, i\neq k}^n W_i}\mathcal{O})_0$, we immediately see that z_j acts bijective on $\mathcal{O}_{I,0}$ if and only if $j \notin I$.

(ii) Let $f(z_1,\dots,z_n) \in (\Gamma_{\prod_{i=1}^n W_i}\mathcal{O})_0$. As $\prod_{i=1}^n W_i$ is simply connected there exists a function

$$F(z_1,\dots,z_n) \in (\Gamma_{\prod_{i=1}^n W_i}\mathcal{O})_0 \text{ such that } \partial_j F = f\,.$$

The other primitives of f w. r. t. ∂_j are given by $F(z_1,\dots,z_n)+C(z_1,\dots,z_{j-1},z_{j+1},\dots,z_n)$ where $C \in (\Gamma_{\prod_{i=1}^n W_i}\mathcal{O})_0$ and C does not depend on z_j. So, in fact $C \in (\Gamma_{\mathbb{C}_j\times\prod_{i=1, i\neq j}^n W_i}\mathcal{O})_0$. Clearly, $j \in I$ if and only if for any such C it follows that $C \in \Lambda_{I,0}$. Now, we see that functions in $\mathcal{O}_{I,0}$ have a uniquely defined primitive w. r. t. ∂_j if and only if $j \in I$ (constants etc. move into the denominator of $\mathcal{O}_{I,0}$). Hence, ∂_j acts bijective on $\mathcal{O}_{I,0}$ if and only if $j \in I$. □

Now, consider "matrix polynomials" of the form z_r^A in $\mathcal{O}_{I,0}$. This is a main ingredient for the solution of the system of equations we are going to be faced with:

Lemma 4.4.
For every $r \in \{1, \dots, n\}$ fix a branch of the logarithm on $\mathbb{C}_r \setminus \mathbb{R}_{\geq 0}$ (recall that $\mathbb{C}^n = \prod_{i=1}^n \mathbb{C}_i$). Let $M \in \mathbb{N}^+$ and let A denote a $M \times M$-matrix with values in $\mathbb{C}$. We set

$$z_r^A := \exp(A \cdot \ln(z_r)) .$$

z_r^A is considered as a matrix with entries in $(\Gamma_{\prod_{i=1}^n W_i}\mathcal{O})_0$ and all entries of z_r^A are invertible w. r. t. multiplication of functions in $(\Gamma_{\prod_{i=1}^n W_i}\mathcal{O})_0$ and $(\Gamma_{\mathbb{C}_t \times \prod_{i=1, i\neq t}^n W_i}\mathcal{O})_0$ for $t \neq r$. Then:

(i) The matrix z_r^A is invertible in $(\Gamma_{\prod_{i=1}^n W_i}\mathcal{O})_0$ and $(\Gamma_{\mathbb{C}_t \times \prod_{i=1, i\neq t}^n W_i}\mathcal{O})_0$ for $t \neq r$.

(ii) Let $I = \{m_1, \dots, m_{|I|}\} \in \mathcal{P}(\{1, \dots, n\})$ and $\{l_1, \dots, l_{n-|I|}\} = \{1, \dots, n\} \setminus I$. Assume we are given pairwise commuting $M \times M$-matrices $A_{m_1}, \dots, A_{m_{|I|}}, A_{l_1}, \dots, A_{l_{n-|I|}}$ with values in $\mathbb{C}$, and the eigenvalues of $A_{m_1}, \dots, A_{m_{|I|}}$ lie in Σ. Let $\lambda = (\lambda_1, \dots, \lambda_M)^T \in \mathbb{C}^M$ and

$$\tilde{\mathcal{F}} := z_{l_1}^{A_{l_1}} \cdot \ldots \cdot z_{l_{n-|I|}}^{A_{l_{n-|I|}}} \cdot z_{m_1}^{A_{m_1}} \cdot \ldots \cdot z_{m_{|I|}}^{A_{m_{|I|}}} .$$

Then:

$$\partial_{m_1}^{-1} \dots \partial_{m_{|I|}}^{-1} \tilde{\mathcal{F}} \cdot \lambda \in (\Lambda_{I,0})^M \iff \lambda = (0, \dots, 0)^T$$

Proof:

(i) Let $\mu_1, \dots, \mu_q \in \mathbb{C}$ denote the eigenvalues of A and let J be the Jordan normal form of A. Obviously, $\exp(J \cdot \ln(z_r))$ is an upper-triangular matrix with $z_r^{\mu_1}, \dots, z_r^{\mu_q}$ on the diagonal. Therefore,

$$\det(\exp(A \cdot \ln(z_r))) = \det(\exp(J \cdot \ln(z_r))) = \prod_{i=1}^{q} (z_r^{\mu_i})^{p_i} \neq 0$$

where $p_1, \dots, p_q \in \mathbb{N}^+$. This yields the invertibility of $\exp(A \cdot \ln(z_r))$ in $(\Gamma_{\prod_{i=1}^n W_i}\mathcal{O})_0$. As $\exp(A \cdot \ln(z_r))$ only depends on the variable z_r we may also consider z_r^A as matrix with entries in $(\Gamma_{\mathbb{C}_t \times \prod_{i=1, i\neq t}^n W_i}\mathcal{O})_0$ for $t \in \{1, \dots, n\} \setminus \{r\}$. Now, with the same arguments as before it follows that z_r^A is invertible in $(\Gamma_{\mathbb{C}_t \times \prod_{i=1, i\neq t}^n W_i}\mathcal{O})_0$ if $t \neq r$.

(ii) We prove "$\Rightarrow$". For simplicity let $I = \{1, \dots, |I|\}$, and assume that we are given $\lambda_1, \dots, \lambda_M$ such that

$$z_{|I|+1}^{A_{|I|+1}} \cdot \ldots \cdot z_n^{A_n} \cdot \partial_1^{-1} \cdot \ldots \cdot \partial_{|I|}^{-1} \cdot z_1^{A_1} \cdot \ldots \cdot z_{|I|}^{A_{|I|}} \cdot \begin{pmatrix} \lambda_1 \\ \vdots \\ \lambda_M \end{pmatrix} \in (\Lambda_{I,0})^M . \tag{$*$}$$

By part (i), $z_{|I|+1}^{A_{|I|+1}}, \dots, z_n^{A_n}$ are invertible and act bijective on $(\Lambda_{I,0})^M$. So, equation $(*)$ is equivalent to

$$\partial_1^{-1} z_1^{A_1} \cdot \ldots \cdot \partial_{|I|}^{-1} z_{|I|}^{A_{|I|}} \cdot \begin{pmatrix} \lambda_1 \\ \vdots \\ \lambda_M \end{pmatrix} \in (\Lambda_{I,0})^M = \sum_{l=1}^{|I|} \left((\Gamma_{\mathbb{C}_l \times \prod_{\substack{i=1 \\ i\neq l}}^n W_i}\mathcal{O})_0 \right)^M .$$

This means, for $l = 1,\ldots,|I|$, we find $f_l(z_1,\ldots,z_n) \in \left((\Gamma_{\mathbb{C}_l\times\prod_{i=1,i\neq l}^n W_i}\mathcal{O})_0\right)^M$ such that

$$\partial_1^{-1} z_1^{A_1}\cdot\ldots\cdot\partial_{|I|}^{-1} z_{|I|}^{A_{|I|}}\cdot\begin{pmatrix}\lambda_1\\ \vdots\\ \lambda_M\end{pmatrix} = \sum_{l=1}^{|I|} f_l(z_1,\ldots,z_n)\,.$$

Let us apply $\mathrm{var}_{|I|}\circ\cdots\circ\mathrm{var}_1$ to both sides of the equation where var_l was given by $M_l - \mathrm{Id}$:

- Let us treat the (LHS): First, let us examine the invertibility of $\mathrm{var}_l(\partial_l^{-1}z_l^{A_l})$: Let J_l denote the Jordan normal form of A_l such that $A_l = B_l^{-1}J_lB_l$ for an invertible matrix B_l. This yields

$$\mathrm{var}_l(\partial_l^{-1}z_l^{A_l}) = \mathrm{var}_l\left(B_l^{-1}(\partial_l^{-1}\exp(J_l\ln(z_l)))B_l\right) = B_l^{-1}\,\mathrm{var}_l\left(\partial_l^{-1}\exp(J_l\ln(z_l))\right)B_l$$

where the second equality follows from the easily verified fact that every entry of $\partial_l^{-1}\exp(J_l\ln(z_l))$ is analytically continuable around $z_l = 0$. As var_l and ∂_l^{-1} act entrywise and the matrix exponential respects the blocks, we consider a single Jordan block. Let $J_{l,a}$ denote the Jordan block with $a\in\Sigma$ on the diagonal and 1 on the first upper off-diagonal. First, assume that $a\neq -1$. Then

$$\begin{aligned}\mathrm{var}_l\left(\partial_l^{-1}\exp(J_{l,a}\ln(z_l))\right) &= \mathrm{var}_l\left((J_{l,a}+\mathrm{Id})^{-1}\exp\left((J_{l,a}+\mathrm{Id})\ln(z_l)\right)\right) = \\ &= (J_{l,a}+\mathrm{Id})^{-1}\cdot\exp((J_{l,a}+\mathrm{Id})\ln(z_l))\cdot\left(e^{2\pi i(J_{l,a}+\mathrm{Id})}-\mathrm{Id}\right).\end{aligned}$$

Using part (i) and $\mathrm{Spec}(J_a+\mathrm{Id})\subset\mathbb{C}\setminus\mathbb{Z}$, this is an invertible matrix in $(\Gamma_{\prod_{i=1}^n W_i}\mathcal{O})_0$. Now, let $a=-1$. The matrix $\exp(J_{l,-1}\ln(z_l))$ is upper-triangular with $\frac{1}{z_l}$ on the diagonal. This yields (up to a matrix which is independent of z_l) that $\partial_l^{-1}\exp(J_{l,-1}\ln(z_l))$ is upper-triangular with $\ln(z_l)$ on the diagonal. Hence, $\mathrm{var}_l\left(\partial_l^{-1}\exp(J_{l,-1}\ln(z_l))\right)$ is an upper-triangular matrix with $2\pi i$ as diagonal entry, and therefore it is invertible in $(\Gamma_{\prod_{i=1}^n W_i}\mathcal{O})_0$. All in all, we see that $\mathrm{var}_l(\partial_l^{-1}z_l^{A_l})$ is invertible in $(\Gamma_{\prod_{i=1}^n W_i}\mathcal{O})_0$. Finally, it is easy to see that

$$(\mathrm{var}_{|I|}\circ\ldots\circ\mathrm{var}_1)(\mathrm{LHS}) = \mathrm{var}_1(\partial_1^{-1}z_1^{A_1})\cdot\ldots\cdot\mathrm{var}_{|I|}(\partial_{|I|}^{-1}z_{|I|}^{A_{|I|}})\cdot\begin{pmatrix}\lambda_1\\ \vdots\\ \lambda_M\end{pmatrix}.$$

- Now, consider the (RHS): As every entry of f_1 is holomorphic in the coordinate z_1 on a small disc around 0, it is in particular analytically continuable around $z_1 = 0$ and $M_1 f_1 = f_1$ as well as $M_1(\sum_{l=1}^{|I|} f_l) = M_1 f_1 + M_1(\sum_{l=2}^{|I|} f_l)$. This yields

$$\begin{aligned}(\mathrm{var}_{|I|}\circ\ldots\circ\mathrm{var}_1)(\mathrm{RHS}) &= (\mathrm{var}_{|I|}\circ\ldots\circ\mathrm{var}_2)\Big(M_1(\sum_{l=1}^{|I|} f_l) - \sum_{l=1}^{|I|} f_l\Big) = \\ &= (\mathrm{var}_{|I|}\circ\ldots\circ\mathrm{var}_2)\Big(M_1(\sum_{l=2}^{|I|} f_l) - \sum_{l=2}^{|I|} f_l\Big) = (\mathrm{var}_{|I|}\circ\ldots\circ\mathrm{var}_1)\Big(\sum_{l=2}^{|I|} f_l\Big).\end{aligned}$$

As the variations commute on the (LHS), we obtain furthermore

$$(\mathrm{var}_{|I|}\circ\ldots\circ\mathrm{var}_1)\Big(\sum_{l=2}^{|I|} f_l\Big) = (\mathrm{var}_{|I|}\circ\ldots\circ\mathrm{var}_3\circ\mathrm{var}_1\circ\mathrm{var}_2)\Big(\sum_{l=2}^{|I|} f_l\Big) =$$

$$= (\mathrm{var}_{|I|} \circ \ldots \circ \mathrm{var}_3 \circ \mathrm{var}_1)\Big(M_2\Big(\sum_{l=2}^{|I|} f_l\Big) - \sum_{l=2}^{|I|} f_l\Big) =$$

$$= (\mathrm{var}_{|I|} \circ \ldots \circ \mathrm{var}_3 \circ \mathrm{var}_1 \circ \mathrm{var}_2)\Big(\sum_{l=3}^{|I|} f_l\Big).$$

Continuing this process, we obtain $(\mathrm{var}_{|I|} \circ \ldots \circ \mathrm{var}_1)(\mathrm{RHS}) = \begin{pmatrix} 0 \\ \vdots \\ 0 \end{pmatrix}$.

This leads to the equality

$$(\mathrm{var}_{|I|} \circ \ldots \circ \mathrm{var}_1)\left(\partial_1^{-1} z_1^{A_1} \cdot \ldots \cdot \partial_{|I|}^{-1} z_{|I|}^{A_{|I|}} \cdot \begin{pmatrix} \lambda_1 \\ \vdots \\ \lambda_M \end{pmatrix}\right) =$$

$$= \mathrm{var}_1(\partial_1^{-1} z_1^{A_1}) \cdot \ldots \cdot \mathrm{var}_{|I|}(\partial_{|I|}^{-1} z_{|I|}^{A_{|I|}}) \cdot \begin{pmatrix} \lambda_1 \\ \vdots \\ \lambda_M \end{pmatrix} = \begin{pmatrix} 0 \\ \vdots \\ 0 \end{pmatrix}.$$

As all matrices $\mathrm{var}_l(\partial_l^{-1} z_l^{A_l})$ are invertible in $(\Gamma_{\prod_{i=1}^n W_i} \mathcal{O})_0$, we obtain $\lambda_1 = \ldots = \lambda_M = 0$ as claimed. □

4.2. Main theorem

Now, we are ready to state and prove the main theorem. It generalises Theorem 3.7 to arbitrary dimension for regular holonomic $\mathscr{D}$-modules whose characteristic variety is contained in Δ_S. As we will see now almost the same categories and functors are involved as in the 1-dimensional case.

Theorem 4.5.
The functors $\mathscr{A} \circ E$ and $Q \circ D$ are naturally isomorphic, i. e. the following diagram commutes up to a natural isomorphism

$$\begin{array}{ccc} \mathcal{M}od^S_{rh}(\mathscr{D}) & \xrightarrow{\mathscr{A} \cong \alpha \circ Sol} & \mathcal{C}_n \\ \uparrow{\scriptstyle E} & & \uparrow{\scriptstyle Q} \\ \mathcal{Q}ui_n^{\Sigma_1} & \xrightarrow{D} & \mathcal{Q}ui_n^{\Sigma_1}. \end{array}$$

In particular, $E\colon \mathcal{Q}ui_n^{\Sigma_1} \to \mathcal{M}od^S_{rh}(\mathscr{D})$ is an equivalence of categories with quasi-inverse $D \circ \mathcal{G} \circ \mathscr{A}$, and $E \circ D \circ \mathcal{G}$ is a quasi-inverse of $\mathscr{A}$.
Furthermore, E is essentially surjective. This means that the category of quiver $\mathscr{D}$-modules is exactly the category $\mathcal{M}od^S_{rh}(\mathscr{D})$, and every $\mathscr{D}$-module in $\mathcal{M}od^S_{rh}(\mathscr{D})$ is in fact isomorphic to a quiver $\mathscr{D}$-module as given in Definition 2.1.

The categories involved in Theorem 4.5 were defined in Definition 1.2 ($\mathcal{C}_n$), Definition 1.3 ($\mathcal{Q}ui_n^{\Sigma_1}$) and Definition 2.3 ($\mathcal{M}od^S_{\mathrm{rh}}(\mathscr{D})$). The functors involved were defined in Definition 1.9 (D), Theorem 1.11 (Q and $\mathcal{G}$), Definition 2.1 and Theorem 2.4 (E), and Theorem 4.2 and the subsequent remark there ($\mathscr{A} \cong \alpha \circ \mathrm{Sol}$).

The proof of the theorem is mainly divided into two propositions. In Proposition 4.6 we examine $\mathrm{Hom}_{\mathscr{D}_{X,0}}((E\mathcal{V}_n)_0, \mathcal{O}_{I,0})$, i. e. how the vector spaces of the object in $\mathcal{C}_n$ look like when applying $\mathscr{A}$ to a quiver $\mathscr{D}$-module. In Proposition 4.8 we analyse how the entire object in $\mathcal{C}_n$ now looks like when applying $\mathscr{A}$ to a quiver $\mathscr{D}$-module. In both propositions we restrict to quiver representations in $\mathcal{Q}ui_n^{\Sigma_1}$. This restriction is mainly necessary because of the non-uniqueness or uniqueness of the logarithm which is part of Corollary 1.13 (and Theorem 1.11) and Lemma 4.4.

Proposition 4.6.
Let $\mathcal{V}_n = (V_I, B_{I\cup\{i\},I}, B_{I,I\cup\{i\}})$ denote an object of the quiver representation category $\mathcal{Q}ui_n^{\Sigma_1}$ and let $E\mathcal{V}_n$ denote the corresponding quiver $\mathscr{D}$-module. Then, we are given a canonical isomorphism

$$\mathfrak{a}\colon V_I^* \xrightarrow{\cong} \mathrm{Hom}_{\mathscr{D}_{X,0}}((E\mathcal{V}_n)_0, \mathcal{O}_{I,0})$$

for every $I \in \mathcal{P}(\{1,\dots,n\})$.

Proof:
We abbreviate $\mathcal{V} = \mathcal{V}_n$. The proof of this lemma will be carried out in several steps:

(i) Let us compute $\mathrm{Hom}_{\mathscr{D}_0}((E\mathcal{V})_0, \mathcal{O}_{I,0})$ using the Definition 2.1 of $E\mathcal{V}$. We have:

$$\begin{aligned}
&\mathrm{Hom}_{\mathscr{D}_0}((E\mathcal{V})_0, \mathcal{O}_{I,0}) = \\
&= \Bigg\{ \phi \in \mathrm{Hom}_{\mathscr{D}_0}\Bigg(\bigoplus_{J\in\mathcal{P}(\{1,\dots,n\})} (\mathscr{D}_0 \otimes \overline{\Omega}_J \otimes V_J)\,,\, \mathcal{O}_{I,0} \Bigg) \,\Bigg|\, \\
&\qquad \phi(\partial_j \otimes \omega_J \otimes v_J - 1 \otimes \omega_{J\cup\{j\}} \otimes B_{J\cup\{j\},J}(v_J)) = 0, \\
&\qquad \phi(z_j \otimes \omega_{J\cup\{j\}} \otimes v_{J\cup\{j\}} - 1 \otimes \omega_J \otimes B_{J,J\cup\{j\}}(v_{J\cup\{j\}})) = 0 \\
&\qquad \text{for } J \neq \{1,\dots,n\},\, j \in \{1,\dots,n\}\setminus J,\, v_J \in V_J,\, v_{J\cup\{j\}} \in V_{J\cup\{j\}} \Bigg\}
\end{aligned}$$

Clearly, the natural isomorphisms

$$\begin{aligned}
&\mathrm{Hom}_{\mathscr{D}_0}\Bigg(\bigoplus_{J\in\mathcal{P}(\{1,\dots,n\})} (\mathscr{D}_0 \otimes \overline{\Omega}_J \otimes V_J), \mathcal{O}_{I,0} \Bigg) \cong \bigoplus_{J\in\mathcal{P}(\{1,\dots,n\})} \mathrm{Hom}_{\mathscr{D}_0}((\mathscr{D}_0 \otimes \overline{\Omega}_J \otimes V_J), \mathcal{O}_{I,0}) \cong \\
&\cong \bigoplus_{J\in\mathcal{P}(\{1,\dots,n\})} \mathrm{Hom}_{\mathbb{C}}(V_J, \mathcal{O}_{I,0})
\end{aligned}$$

hold. These yield that

$$\begin{aligned}
&\mathrm{Hom}_{\mathscr{D}_0}((E\mathcal{V})_0, \mathcal{O}_{I,0}) \cong \\
&\cong \Bigg\{ \bigoplus_{J\in\mathcal{P}(\{1,\dots,n\})} \phi^J \in \bigoplus_{J\in\mathcal{P}(\{1,\dots,n\})} \mathrm{Hom}_{\mathbb{C}}(V_J, \mathcal{O}_{I,0}) \,\Bigg|\, \\
&\qquad \partial_j \cdot \phi^J(v_J) - \phi^{J\cup\{j\}}(B_{J\cup\{j\},J}(v_J)) = 0, \\
&\qquad z_j \cdot \phi^{J\cup\{j\}}(v_{J\cup\{j\}}) - \phi^J(B_{J,J\cup\{j\}}(v_{J\cup\{j\}})) = 0 \\
&\qquad \text{for } J \neq \{1,\dots,n\},\, j \in \{1,\dots,n\}\setminus J,\, v_J \in V_J,\, v_{J\cup\{j\}} \in V_{J\cup\{j\}} \Bigg\}.
\end{aligned}$$

(ii) Consider the following system of equations from step (i):

$$\begin{aligned} \partial_j \cdot \phi^J - \phi^{J\cup\{j\}} \circ B_{J\cup\{j\},J} &= 0 \\ z_j \cdot \phi^{J\cup\{j\}} - \phi^J \circ B_{J,J\cup\{j\}} &= 0 \end{aligned} \tag{$\tilde{\star}$}$$

where $\phi^J \in \mathrm{Hom}_{\mathbb{C}}(V_J, \mathcal{O}_{I,0})$ and $J \neq \{1,\dots,n\}$, $j \in \{1,\dots,n\} \setminus J$.

Let us use Lemma 4.3: The bijectivity of z_j for $j \notin I$ and of ∂_j for $j \in I$ acting on $\mathcal{O}_{I,0}$ allows us to express step by step every ϕ^K uniquely in terms of ϕ^I for $K \in \mathcal{P}(\{1,\dots,n\})\setminus I$. Therefor the following algorithm (ALG) may be used:

Fix $K \in \mathcal{P}(\{1,\dots,n\}) \setminus I$. We can express $K \cup I$ as the union of the three disjoint sets $K_1 := K \cap I$, $K_2 := K \setminus K_1$, $K_3 := I \setminus K_1$. At least one of the sets K_2 and K_3 is non-empty as $K \neq I$. We note again that z_l, ∂_m act bijective on $\mathcal{O}_{I,0}$ for $l \in K_2$ and $m \in K_3$.

1. Step: If K_2 is empty, i. e. $K = K_1$, skip this first step. Otherwise proceed as follows: For $l_1 \in K_2$ use the equation

$$z_{l_1} \cdot \phi^K - \phi^{K\setminus\{l_1\}} \circ B_{K\setminus\{l_1\},K} = 0$$

to obtain

$$\phi^K = \frac{1}{z_{l_1}} \cdot \left(\phi^{K\setminus\{l_1\}} \circ B_{K\setminus\{l_1\},K}\right).$$

For $l_2 \in K_2 \setminus \{l_1\}$ use the equation

$$z_{l_2} \cdot \phi^{K\setminus\{l_1\}} - \phi^{K\setminus\{l_1,l_2\}} \circ B_{K\setminus\{l_1,l_2\},K\setminus\{l_1\}} = 0$$

to express ϕ^K in terms of $\phi^{K\setminus\{l_1,l_2\}}$. Continue until ϕ^K is expressed in terms of ϕ^{K_1}.

2. Step: If K_3 is empty, then stop here. In this case $K_1 = I$, i. e. we already expressed ϕ^K in terms of ϕ^I. If K_3 is non-empty, proceed as follows: For $m_1 \in K_3$ use the equation

$$\partial_{m_1} \cdot \phi^{K_1} - \phi^{K_1\cup\{m_1\}} \circ B_{K_1\cup\{m_1\},K_1} = 0$$

which yields

$$\phi^{K_1} = \partial_{m_1}^{-1} \cdot \left(\phi^{K_1\cup\{m_1\}} \circ B_{K_1\cup\{m_1\},K_1}\right).$$

For $m_2 \in K_3 \setminus \{m_1\}$ use the equation

$$\partial_{m_2} \cdot \phi^{K_1\cup\{m_1\}} - \phi^{K_1\cup\{m_1,m_2\}} \circ B_{K_1\cup\{m_1,m_2\},K_1\cup\{m_1\}} = 0$$

to express ϕ^{K_1} in terms of $\phi^{K_1\cup\{m_1,m_2\}}$. Continue this process until ϕ^{K_1} – and therefore ϕ^K – is expressed in terms of ϕ^I.

Note that the order in which we solve for ϕ^I in the algorithm (ALG) does not influence the result. This is ensured by the commutativity conditions on the maps $B_{\bullet,\bullet}$ (see Definition 1.1) and the fact that z_l commutes with ∂_m for $l \notin I$ and $m \in I$. Thus, any other manner solving for ϕ^I which uses the equations involving z_l and ∂_m for $l \notin I$, $m \in I$ (the equations with bijective z_l and ∂_m) will produce the same result. This shows us that every ϕ^K can even be uniquely expressed in terms of ϕ^I.

Using (ALG), we see that ($\widetilde{\star}$) implies that $\phi^I \in \mathrm{Hom}_{\mathbb{C}}(V_I, \mathcal{O}_{I,0})$ fulfils the following system ($\star$) of n equations

$$\left\{\begin{array}{r} \partial_l \cdot \phi^I - \frac{1}{z_l} \cdot \left(\phi^I \circ B_{I,I\cup\{l\}} \circ B_{I\cup\{l\},I}\right) = 0 \\ z_m \cdot \phi^I - \partial_m^{-1} \cdot \left(\phi^I \circ B_{I,I\setminus\{m\}} \circ B_{I\setminus\{m\},I}\right) = 0 \end{array}\right\} \Longleftrightarrow$$

$$\left\{\begin{array}{r} z_l\partial_l \cdot \phi^I - \phi^I \circ (B_{I,I\cup\{l\}} B_{I\cup\{l\},I}) = 0 \\ z_m\partial_m \cdot \phi^I - \phi^I \circ (B_{I,I\setminus\{m\}} B_{I\setminus\{m\},I} - \mathrm{Id}) = 0 \end{array}\right\} \tag{$\star$}$$

where $l \notin I$, $m \in I$.

On the other hand ($\widetilde{\star}$) is likewise implied by ($\star$):
Assume that we are given $\phi^I \in \mathrm{Hom}_{\mathbb{C}}(V_I, \mathcal{O}_{I,0})$ solving ($\star$). Then use (ALG) as definition of ϕ^K for all $K \in \mathcal{P}(\{1,\ldots,n\}) \setminus I$. Consider a pair of equations from ($\widetilde{\star}$) for a fixed set $J \in \mathcal{P}(\{1,\ldots,n\})$, $J \neq \{1,\ldots,n\}$

$$\begin{aligned} \partial_j \cdot \phi^J - \phi^{J\cup\{j\}} \circ B_{J\cup\{j\},J} &= 0 \\ z_j \cdot \phi^{J\cup\{j\}} - \phi^J \circ B_{J,J\cup\{j\}} &= 0 \end{aligned}$$

where $j \in \{1,\ldots,n\} \setminus J$. If $j \notin I$, the second equation is clearly fulfilled as (ALG) gives $\phi^{J\cup\{j\}} = z_j^{-1} \cdot (\phi^J \circ B_{J,J\cup\{j\}})$. We plug this and the expression for ϕ^J in terms of ϕ^I into the first equation. This yields

$$z_{k_1}^{-1} \cdot \ldots \cdot z_{k_{|J\setminus I|}}^{-1} \cdot \partial_{p_1}^{-1} \cdot \ldots \cdot \partial_{p_{|I\setminus J|}}^{-1} \cdot \left(z_j\partial_j \cdot \phi^I - \phi^I \circ (B_{I,I\cup\{j\}} B_{I\cup\{j\},I})\right) \circ A = 0$$

where $\{k_1,\ldots,k_{|J\setminus I|}\} = J \setminus I$, $\{p_1,\ldots,p_{|I\setminus J|}\} = I \setminus J$ and $A \in \mathrm{Hom}_{\mathbb{C}}(V_J, V_I)$. Clearly, ($\star$) implies this equation. That the equation is indeed of this form becomes clear by imagining the representation of the quiver being placed on a hypercube as described at the beginning of Chapter 1 and using the commutativity of the maps $B_{\bullet,\bullet}$.
Similarly it works if $j \in I$: Then, we can trace back the pair of equations to the equation $z_j\partial_j \cdot \phi^I - \phi^I \circ (B_{I,I\setminus\{j\}} B_{I\setminus\{j\},I} - \mathrm{Id}) = 0$. All in all, we see that ($\star$) and ($\widetilde{\star}$) are equivalent.

This shows us that in fact all ϕ^J can be uniquely recovered from ϕ^I and we have

$$\begin{aligned} &\mathrm{Hom}_{\mathscr{D}_0}((E\mathcal{V})_0, \mathcal{O}_{I,0}) \cong \\ &\cong \Big\{ \phi^I \in \mathrm{Hom}_{\mathbb{C}}(V_I, \mathcal{O}_{I,0}) \;\Big|\; z_l\partial_l \cdot \phi^I(v_I) - \phi^I(B_{I,I\cup\{l\}} B_{I\cup\{l\},I}(v_I)) = 0\,, \\ &\qquad z_m\partial_m \cdot \phi^I(v_I) - \phi^I((B_{I,I\setminus\{m\}} B_{I\setminus\{m\},I} - \mathrm{Id})(v_I)) = 0 \\ &\qquad \text{for } l \in \{1,\ldots,n\} \setminus I,\ m \in I,\ v_I \in V_I \Big\}\,. \end{aligned} \tag{1}$$

(iii) If $V_I = 0$, we immediately obtain $\phi^I = 0$. Hence, $\phi^K = 0$ for all $K \in \mathcal{P}(\{1,\ldots,n\})$ by part (ii). This gives us

$$\mathrm{Hom}_{\mathscr{D}_0}((E\mathcal{V})_0, \mathcal{O}_{I,0}) = 0\,.$$

Thus, for the rest of the proof we assume that $\dim_{\mathbb{C}}(V_I) \geq 1$.

(iv) The dimension of $\mathrm{Hom}_{\mathscr{D}_0}((E\mathcal{V})_0, \mathcal{O}_{I,0}))$ over $\mathbb{C}$ must be finite (see [GGM85b]). This follows mainly from Kashiwara's constructability theorem (see for instance [HTT08, Section 4.6]).

We use the following proposition of [GGM85b] to give an upper bound of the dimension of $\operatorname{Hom}_{\mathscr{D}_0}((E\mathcal{V})_0, \mathcal{O}_{I,0})$:

> Let $z_I^* = (z_1, \dots, z_n, \xi_1, \dots, \xi_n) \in T^*\mathbb{C}^n$ verifying $z_i \xi_i = 0$ for all i, and $z_i = 0 \Leftrightarrow i \in I$ and $\xi_i \neq 0 \Leftrightarrow i \in I$. Then
>
> $$\dim_{\mathbb{C}} \operatorname{Hom}_{\mathscr{D}_0}((E\mathcal{V})_0, \mathcal{O}_{I,0}) = \operatorname{mult}_{z_I^*} E\mathcal{V}\,.$$

We use the definition of the multiplicity of a $\mathscr{D}_X$-module at a certain point of T^*X as given in [GM93, Chapter V]. Supplementary, we will use [Ser75, Subsection II.B)4]. As the definition becomes clear during the following computations, we do not repeat it here.

Let us use the same good filtration on $E\mathcal{V}$ as in the proof of Theorem 2.4. Its sections over $U \subset \mathbb{C}^n$, open, are given by

$$F_k E\mathcal{V}(U) = \frac{F_k\mathscr{D}(U) \otimes \left(\bigoplus_J \overline{\Omega}_J \otimes V_J\right)}{\left(F_k\mathscr{D}(U) \otimes (\bigoplus_J \overline{\Omega}_J \otimes V_J)\right) \cap \mathcal{J}(U)}$$

for $k \in \mathbb{N}_0$, and for $k \in \mathbb{Z} \setminus \mathbb{N}_0$ we have $F_k E\mathcal{V} = 0$. The sheaf $\mathcal{J}$ was given in Definition 2.1. We set

$$\operatorname{gr}_k^F E\mathcal{V} := F_k E\mathcal{V} \big/ F_{k-1} E\mathcal{V} \quad \text{and} \quad \operatorname{gr}^F E\mathcal{V} := \bigoplus_{k \in \mathbb{N}_0} \operatorname{gr}_k^F E\mathcal{V}\,.$$

Let $k \in \mathbb{N}_0$. Let us fix a point $\tilde{z}_I^* = (\tilde{z}_1, \dots, \tilde{z}_n, \tilde{\xi}_1, \dots, \tilde{\xi}_n) =: (\tilde{z}_I, \tilde{\xi}_I)$ where $\tilde{z}_i = 0 \Leftrightarrow i \in I$ and $\tilde{\xi}_i \neq 0 \Leftrightarrow i \in I$. Set

$$M := (\operatorname{gr}^F E\mathcal{V})_{\tilde{z}_I}\,.$$

The exactness of taking stalks gives us

$$M = \bigoplus_{k \in \mathbb{N}_0} (F_k E\mathcal{V})_{\tilde{z}_I} \big/ (F_{k-1} E\mathcal{V})_{\tilde{z}_I} \quad \text{and} \quad (F_k E\mathcal{V})_{\tilde{z}_I} = \frac{F_k\mathscr{D}_{\tilde{z}_I} \otimes \left(\bigoplus_J \overline{\Omega}_J \otimes V_J\right)}{\left(F_k\mathscr{D}_{\tilde{z}_I} \otimes (\bigoplus_J \overline{\Omega}_J \otimes V_J)\right) \cap \mathcal{J}_{\tilde{z}_I}}\,.$$

z_i is an invertible element in $F_k\mathscr{D}_{\tilde{z}_I}$ if and only if $i \notin I$. So let $i \in \{1, \dots, n\} \setminus I$. Furthermore, let $K \in \mathcal{P}(\{1, \dots, n\} \setminus \{i\})$. We denote by $[P \otimes \omega_{K\cup\{i\}} \otimes v_{K\cup\{i\}}]$ the image of

$$P \otimes \omega_{K\cup\{i\}} \otimes v_{K\cup\{i\}} \in F_k\mathscr{D}_{\tilde{z}_I} \otimes \overline{\Omega}_{K\cup\{i\}} \otimes V_{K\cup\{i\}}$$

in $(F_k E\mathcal{V})_{\tilde{z}_I}$. We have the following identity in $(F_k E\mathcal{V})_{\tilde{z}_I}$:

$$[P \otimes \omega_{K\cup\{i\}} \otimes v_{K\cup\{i\}}] = [z_i^{-1} P \otimes \omega_K \otimes B_{K,K\cup\{i\}}(v_{K\cup\{i\}})]$$

This allows us to "eliminate" all summands $[F_k\mathscr{D}_{\tilde{z}_I} \otimes \overline{\Omega}_J \otimes V_J]$ in $(F_k E\mathcal{V})_{\tilde{z}_I}$ with $J \setminus I \neq \varnothing$. Hence, we may assume without loss of generality that

$$(F_k E\mathcal{V})_{\tilde{z}_I} = \frac{F_k\mathscr{D}_{\tilde{z}_I} \otimes \left(\bigoplus_{J\subseteq I} \overline{\Omega}_J \otimes V_J\right)}{\left(F_k\mathscr{D}_{\tilde{z}_I} \otimes (\bigoplus_{J\subseteq I} \overline{\Omega}_J \otimes V_J)\right) \cap \mathcal{J}_{\tilde{z}_I}}$$

or likewise that

$$E\mathcal{V} = \frac{\mathscr{D} \otimes \left(\bigoplus_{J\subseteq I} \overline{\Omega}_J \otimes V_J\right)}{\left(\mathscr{D} \otimes (\bigoplus_{J\subseteq I} \overline{\Omega}_J \otimes V_J)\right) \cap \mathcal{J}}\,.$$

Let $\mathcal{O}_{\tilde{z}_I}[\xi_1, \dots, \xi_n] \cong \mathrm{gr}^F \mathscr{D}_{\tilde{z}_I}$ denote the ring of polynomials in n variables with coefficients in $\mathcal{O}_{\tilde{z}_I}$. M is a finitely generated $\mathcal{O}_{\tilde{z}_I}[\xi_1, \dots, \xi_n]$-module. We denote by $\mathcal{M}ax$ the maximal ideal of the local ring $\mathcal{O}_{\tilde{z}_I}$. Let

$$Q_{\tilde{\xi}_I} := \mathcal{M}ax + (\xi_1 - \tilde{\xi}_1, \dots, \xi_n - \tilde{\xi}_n).$$

This defines a maximal ideal in $\mathcal{O}_{\tilde{z}_I}[\xi_1, \dots, \xi_n]$ as $\mathcal{O}_{\tilde{z}_I}[\xi_1, \dots, \xi_n]/Q_{\tilde{\xi}_I} \cong \mathcal{O}_{\tilde{z}_I}/\mathcal{M}ax$ is a field. Thus, $M/Q_{\tilde{\xi}_I}M$ is a finitely generated $\mathcal{O}_{\tilde{z}_I}[\xi_1, \dots, \xi_n]/Q_{\tilde{\xi}_I}$-vector space. In particular $M/Q_{\tilde{\xi}_I}M$ has finite length over $\mathcal{O}_{\tilde{z}_I}[\xi_1, \dots, \xi_n]/Q_{\tilde{\xi}_I}$. Therefore, there exists a polynomial $P_{M,Q_{\tilde{\xi}_I}}(N)$, called Hilbert-Samuel polynomial, and an integer $N_0 \in \mathbb{N}$ such that

$$P_{M,Q_{\tilde{\xi}_I}}(N) = \mathrm{length}(M/Q_{\tilde{\xi}_I}^N M) \quad \text{for all } N \geq N_0.$$

The highest degree term of P has the form[1]

$$\frac{e}{d!}N^d \quad \text{where } e \in \mathbb{N},\ d \in \mathbb{N}$$

and by definition

$$e = \mathrm{mult}_{\tilde{z}_I^*} E\mathcal{V}.$$

Let us use [Ser75, Proposition 11a) in Subsection II.B)4]: We obtain

$$P_{M,Q_{\tilde{\xi}_I}}(N) = \sum_{T=\mathcal{O}_{\tilde{z}_I}[\xi_1,\dots,\xi_n]\setminus\mathfrak{m}} P_{T^{-1}M,T^{-1}Q_{\tilde{\xi}_I}}(N) = \sum_{\substack{T=\mathcal{O}_{\tilde{z}_I}[\xi_1,\dots,\xi_n]\setminus\mathfrak{m} \\ Q_{\tilde{\xi}_I}\subset\mathfrak{m}}} P_{T^{-1}M,T^{-1}Q_{\tilde{\xi}_I}}(N)$$

where $\mathfrak{m}$ runs over all maximal ideals of $\mathcal{O}_{\tilde{z}_I}[\xi_1, \dots, \xi_n]$. But as $Q_{\tilde{\xi}_I}$ is an maximal ideal as well, we obtain

$$P_{M,Q_{\tilde{\xi}_I}}(N) = P_{T^{-1}M,T^{-1}Q_{\tilde{\xi}_I}}(N) \quad \text{for } T := \mathcal{O}_{\tilde{z}_I}[\xi_1, \dots, \xi_n] \setminus Q_{\tilde{\xi}_I}.$$

So we need to consider the localisation of M at T:

$$T^{-1}M = \bigoplus_{k\in\mathbb{N}_0} T^{-1}(\mathrm{gr}_k^F E\mathcal{V})_{\tilde{z}_I}$$

Let $[P \otimes \omega_K \otimes v_K]$ denote the image of $P \otimes \omega_K \otimes v_K \in F_k\mathscr{D}_{\tilde{z}_I} \otimes \overline{\Omega}_K \otimes V_K$ in $(\mathrm{gr}_k^F E\mathcal{V})_{\tilde{z}_I}$ for $K \subsetneq I$. For every $i \in I \setminus K$ we have the following identity in $(\mathrm{gr}_{k+1}^F E\mathcal{V})_{\tilde{z}_I}$:

$$\xi_i \cdot [P \otimes \omega_K \otimes v_K] = [P \otimes \omega_{K\cup\{i\}} \otimes B_{K\cup\{i\},K}(v_K)] = 0$$

Consider the same identity in $T^{-1}M$. The element ξ_i is invertible in $T^{-1}M$ as $\tilde{\xi}_i \neq 0$ for $i \in I$. This means we may consider $\xi_i \cdot \_ : T^{-1}(\mathrm{gr}_k^F E\mathcal{V})_{\tilde{z}_I} \to T^{-1}(\mathrm{gr}_{k+1}^F E\mathcal{V})_{\tilde{z}_I}$ as bijective map. Therefore,

$$\xi_i \cdot \frac{[P \otimes \omega_K \otimes v_K]}{1} = 0 \quad \text{yields} \quad \frac{[P \otimes \omega_K \otimes v_K]}{1} = 0.$$

[1] We note that [GM93] are using $\mathrm{length}(M/Q_{\tilde{\xi}_I}^{N+1}M)$ instead of $\mathrm{length}(M/Q_{\tilde{\xi}_I}^N M)$ to define P. However, this does not affect the highest degree term of P.

Thus, we may assume without loss of generality that

$$T^{-1}(F_k E\mathcal{V})_{\tilde{z}_I} = \frac{T^{-1}F_k\mathscr{D}_{\tilde{z}_I} \otimes \overline{\Omega}_I \otimes V_I}{\left(T^{-1}F_k\mathscr{D}_{\tilde{z}_I} \otimes \overline{\Omega}_I \otimes V_I\right) \cap T^{-1}\mathcal{J}_{\tilde{z}_I}}.$$

Using [Ser75, Proposition 11a) in Subsection II.B)4] the other way round, we may assume without loss of generality that

$$(F_k E\mathcal{V})_{\tilde{z}_I} = \frac{F_k\mathscr{D}_{\tilde{z}_I} \otimes \overline{\Omega}_I \otimes V_I}{\left(F_k\mathscr{D}_{\tilde{z}_I} \otimes \overline{\Omega}_I \otimes V_I\right) \cap \mathcal{J}_{\tilde{z}_I}}$$

or likewise that

$$E\mathcal{V} = \frac{\mathscr{D} \otimes \overline{\Omega}_I \otimes V_I}{\left(\mathscr{D} \otimes \overline{\Omega}_I \otimes V_I\right) \cap \mathcal{J}}.$$

Consider the following exact sequence of holonomic $\mathscr{D}$-modules

$$0 \longrightarrow \ker(\pi) \hookrightarrow \widetilde{\mathscr{N}} \overset{\pi}{\twoheadrightarrow} E\mathcal{V} \longrightarrow 0$$

where

$$\widetilde{\mathscr{N}} := \mathscr{D} \otimes V_I \Big/ \left(z_{m_1} \otimes V_I, \ldots, z_{m_{|I|}} \otimes V_I, \partial_{l_1} \otimes V_I, \ldots, \partial_{l_{n-|I|}} \otimes V_I\right)$$

and $\{m_1, \ldots, m_{|I|}\} = I$, $\{l_1, \ldots, l_{n-|I|}\} = \{1, \ldots, n\} \setminus I$. This exact sequence yields now $\operatorname{mult}_{\tilde{z}_I^*} \widetilde{\mathscr{N}} = \operatorname{mult}_{\tilde{z}_I^*} E\mathcal{V} + \operatorname{mult}_{\tilde{z}_I^*} \ker(\pi)$ and therefore

$$\operatorname{mult}_{\tilde{z}_I^*} E\mathcal{V} \leq \operatorname{mult}_{\tilde{z}_I^*} \widetilde{\mathscr{N}}.$$

Moreover, it is easy to see that

$$\widetilde{\mathscr{N}} \cong \bigoplus_{n_I\text{-times}} \underbrace{\mathscr{D} \Big/ \left(z_{m_1}, \ldots, z_{m_{|I|}}, \partial_{l_1}, \ldots, \partial_{l_{n-|I|}}\right)}_{=:\mathscr{N}}$$

where $n_I := \dim_{\mathbb{C}}(V_I)$. Using exact sequences as the one above, we obtain step by step that

$$\operatorname{mult}_{\tilde{z}_I^*} \widetilde{\mathscr{N}} = n_I \cdot \operatorname{mult}_{\tilde{z}_I^*} \mathscr{N}.$$

So let us compute $\operatorname{mult}_{\tilde{z}_I^*} \mathscr{N}$: First for simplicity, we reorder the variables such that

$$\mathscr{N} = \mathscr{D} \Big/ \left(z_1, \ldots, z_{|I|}, \partial_{|I|+1}, \ldots, \partial_n\right).$$

This means we consider the multiplicity of $\mathscr{N}$ at

$$\tilde{z}_I^* = (0, \ldots, 0, \tilde{z}_{|I|+1}, \ldots, \tilde{z}_n, \tilde{\xi}_1, \ldots, \tilde{\xi}_{|I|}, 0, \ldots, 0) =: (\tilde{z}_I, \tilde{\xi}_I)$$

where $\tilde{z}_{|I|+1}, \ldots, \tilde{z}_n, \tilde{\xi}_1, \ldots, \tilde{\xi}_{|I|} \neq 0$. We use the good filtration $F_\bullet\mathscr{N}$ on $\mathscr{N}$ which is induced by the standard filtration $F_\bullet\mathscr{D}$ of $\mathscr{D}$. This yields

$$(\operatorname{gr}^F \mathscr{N})_{\tilde{z}_I} \cong \mathbb{C}\{z_{|I|+1} - \tilde{z}_{|I|+1}, \ldots, z_n - \tilde{z}_n\}[\xi_1, \ldots, \xi_{|I|}].$$

Consider $(\operatorname{gr}_F \mathscr{N})_{\tilde{z}_I}$ as a module over

$$(\operatorname{gr}^F \mathscr{D})_{\tilde{z}_I} \cong \mathbb{C}\{z_1, \ldots, z_{|I|}, z_{|I|+1} - \tilde{z}_{|I|+1}, \ldots, z_n - \tilde{z}_n\}[\xi_1, \ldots, \xi_n].$$

Let $\mathcal{M}ax$ be the maximal ideal of $\mathbb{C}\{z_1,\dots,z_{|I|},z_{|I|+1}-\tilde{z}_{|I|+1},\dots,z_n-\tilde{z}_n\}$. We need to compute the multiplicity of $(\mathrm{gr}^F\mathscr{N})_{\tilde{z}_I}$ with respect to the maximal ideal

$$\mathcal{M}ax+(\xi_1-\tilde{\xi}_1,\dots,\xi_{|I|}-\tilde{\xi}_{|I|},\xi_{|I|+1},\dots,\xi_n)$$

of $(\mathrm{gr}^F\mathscr{D})_{\tilde{z}_I}$. A shift of coordinates gives us that we equivalently have to treat

$$\mathbb{C}\{z_{|I|+1},\dots,z_n\}[\xi_1,\dots,\xi_{|I|}]$$

as a module over

$$\mathbb{C}\{z_1,\dots,z_n\}[\xi_1,\dots,\xi_n],$$

and compute its multiplicity with respect to the maximal ideal

$$Q:=(z_1,\dots,z_n,\xi_1-\tilde{\xi}_1,\dots,\xi_{|I|}-\tilde{\xi}_{|I|},\xi_{|I|+1},\dots,\xi_n).$$

This means we have to compute

$$\begin{aligned}&\mathrm{length}\left(\frac{\mathbb{C}\{z_{|I|+1},\dots,z_n\}[\xi_1,\dots,\xi_{|I|}]}{Q^N\cdot\mathbb{C}\{z_{|I|+1},\dots,z_n\}[\xi_1,\dots,\xi_{|I|}]}\right)=\\&=\mathrm{length}\left(\frac{\mathbb{C}\{z_{|I|+1},\dots,z_n\}[\xi_1,\dots,\xi_{|I|}]}{(z_{|I|+1},\dots,z_n,\xi_1-\tilde{\xi}_1,\dots,\xi_{|I|}-\tilde{\xi}_{|I|})^N\cdot\mathbb{C}\{z_{|I|+1},\dots,z_n\}[\xi_1,\dots,\xi_{|I|}]}\right).\end{aligned}$$

But this is exactly equal to the number of monomials of degree less than N in $\mathbb{C}\{z_{|I|+1},\dots,z_n\}[\xi_1,\dots,\xi_{|I|}]$ which is equal to

$$\binom{N-1+n}{N-1}=1\cdot\frac{N^n}{n!}+\text{polynomial in } N \text{ of degree less than } n\,.$$

This shows us that $\mathrm{mult}_{\tilde{z}_I^*}\mathscr{N}=1$ and $\mathrm{mult}_{\tilde{z}_I^*}\mathcal{EV}\le n_I$.

(v) Now, we construct the canonical isomorphism η_I from V_I^* into (1). For this purpose let $\alpha\in V_I^*$. We define $\eta_I(\alpha)$ as follows:

Let $\{m_1,\dots,m_{|I|}\}=I$, $\{l_1,\dots,l_{n-|I|}\}=\{1,\dots,n\}\setminus I$ as before. For a moment fix a basis of V_I and denote it by $v_{I,1},\dots,v_{I,n_I}$. Let us denote the matrices corresponding to the maps

$$\alpha\quad\text{and}\quad\mathcal{B}_{I,I\cup\{l\}}:=B_{I,I\cup\{l\}}B_{I\cup\{l\},I}\,,\quad\mathcal{B}_{I,I\setminus\{m\}}:=B_{I,I\setminus\{m\}}B_{I\setminus\{m\},I}$$

w. r. t. this basis by the same symbols in abuse of notation. For every $r\in\{1,\dots,n\}$ fix a branch of the logarithm defined on $\mathbb{C}_r\setminus\mathbb{R}_{\ge0}$. For a square matrix A we already defined $z_r^A=\exp(A\cdot\ln(z_r))$ which fulfils

$$\partial_r z_r^A=A\cdot\tfrac{1}{z_r}\cdot z_r^A=A\cdot z_r^{A-\mathrm{Id}}\quad\text{and}\quad z_r\partial_r z_r^A=A\cdot z_r^A\,.$$

We set

$$\begin{aligned}&\mathscr{F}:=z_{l_1}^{\mathcal{B}_{I,I\cup\{l_1\}}}\cdot\ldots\cdot z_{l_{n-|I|}}^{\mathcal{B}_{I,I\cup\{l_{n-|I|}\}}}\cdot z_{m_1}^{\mathcal{B}_{I,I\setminus\{m_1\}}-\mathrm{Id}}\cdot\ldots\cdot z_{m_{|I|}}^{\mathcal{B}_{I,I\setminus\{m_{|I|}\}}-\mathrm{Id}}\quad\text{and}\\&\eta_I(\alpha):=\alpha\cdot\mathscr{F}\,.\end{aligned}$$

It is easily verified that $\eta_I(\alpha)$ is indeed an element in (1). One only needs to plug $\eta_I(\alpha)$ into ($\star$), and use the fact that $\mathcal{B}_{I,J_1}$ and $\mathcal{B}_{I,J_2}$ commute for all $J_1,J_2\in\mathcal{P}(\{1,\dots,n\})$ that are adjacent to I, i. e. $I=J\cup\{j\}$ or $J=I\cup\{i\}$ for some $i,j\in\{1,\dots,n\}$.

We need to verify that this construction of η_I is independent of the choice of basis of V_I. So, let $\tilde{v}_{I,1}, \dots, \tilde{v}_{I,n_I}$ denote another basis of V_I. Let $\tilde{\mathcal{B}}_{I,I\cup\{l\}}$, $\tilde{\mathcal{B}}_{I,I\setminus\{m\}}$ and $\tilde{\alpha}$ denote the matrices corresponding to the linear maps $\mathcal{B}_{I,I\cup\{l\}}$, $\mathcal{B}_{I,I\setminus\{m\}}$ and α w. r. t. this new basis. Let R denote the matrix of the change of coordinates from $\{v_{I,1}, \dots, v_{I,n_I}\}$ to $\{\tilde{v}_{I,1}, \dots, \tilde{v}_{I,n_I}\}$. This means

$$\mathcal{B}_{I,I\cup\{l\}} = R^{-1}\tilde{\mathcal{B}}_{I,I\cup\{l\}}R\,, \quad \mathcal{B}_{I,I\setminus\{m\}} = R^{-1}\tilde{\mathcal{B}}_{I,I\setminus\{m\}}R \quad \text{and} \quad \alpha = \tilde{\alpha}R\,.$$

Let $v_I \in V_I$. We denote by v_I in abuse of notation the vector w. r. t. the basis $\{v_{I,1}, \dots, v_{I,n_I}\}$ and by $\tilde{v}_I$ the vector w. r. t. the basis $\{\tilde{v}_{I,1}, \dots, \tilde{v}_{I,n_I}\}$. Hence $v_I = R^{-1}\tilde{v}_I$. We obtain

$$\begin{aligned}
\eta_I(\alpha)(v_I) &= \alpha \cdot z_{l_1}^{\mathcal{B}_{I,I\cup\{l_1\}}} \cdot \ldots \cdot z_{l_{n-|I|}}^{\mathcal{B}_{I,I\cup\{l_{n-|I|}\}}} \cdot z_{m_1}^{\mathcal{B}_{I,I\setminus\{m_1\}}-\mathrm{Id}} \cdot \ldots \cdot z_{m_{|I|}}^{\mathcal{B}_{I,I\setminus\{m_{|I|}\}}-\mathrm{Id}} \cdot v_I = \\
&= \tilde{\alpha}RR^{-1}z_{l_1}^{\tilde{\mathcal{B}}_{I,I\cup\{l_1\}}}R \ldots R^{-1}z_{l_{n-|I|}}^{\tilde{\mathcal{B}}_{I,I\cup\{l_{n-|I|}\}}} z_{m_1}^{\tilde{\mathcal{B}}_{I,I\setminus\{m_1\}}-\mathrm{Id}}R \ldots R^{-1}z_{m_{|I|}}^{\tilde{\mathcal{B}}_{I,I\setminus\{m_{|I|}\}}-\mathrm{Id}}RR^{-1}\tilde{v}_I = \\
&= \eta_I(\tilde{\alpha})(\tilde{v}_I)\,.
\end{aligned}$$

Hence, our construction is independent of the choice of basis of V_I.

Now, we want to check that η_I is injective. So assume that $\eta_I(\alpha)$ is the zero mapping. This is equivalent to

$$z_{l_1}^{\mathcal{B}^T_{I,I\cup\{l_1\}}} \cdot \ldots \cdot z_{l_{n-|I|}}^{\mathcal{B}^T_{I,I\cup\{l_{n-|I|}\}}} \cdot z_{m_1}^{\mathcal{B}^T_{I,I\setminus\{m_1\}}-\mathrm{Id}} \cdot \ldots \cdot z_{m_{|I|}}^{\mathcal{B}^T_{I,I\setminus\{m_{|I|}\}}-\mathrm{Id}} \cdot \alpha^T = \begin{pmatrix} 0 \\ \vdots \\ 0 \end{pmatrix}.$$

As ∂_m acts bijective on $\mathcal{O}_{I,0}$ for $m \in I$ (see Lemma 4.3), this is equivalent to

$$\partial_{m_1}^{-1} \ldots \partial_{m_{|I|}}^{-1} z_{l_1}^{\mathcal{B}^T_{I,I\cup\{l_1\}}} \cdot \ldots \cdot z_{l_{n-|I|}}^{\mathcal{B}^T_{I,I\cup\{l_{n-|I|}\}}} \cdot z_{m_1}^{\mathcal{B}^T_{I,I\setminus\{m_1\}}-\mathrm{Id}} \cdot \ldots \cdot z_{m_{|I|}}^{\mathcal{B}^T_{I,I\setminus\{m_{|I|}\}}-\mathrm{Id}} \cdot \alpha^T = \begin{pmatrix} 0 \\ \vdots \\ 0 \end{pmatrix}.$$

Using the definition of $\mathcal{O}_{I,0}$, we equivalently have

$$\partial_{m_1}^{-1} \ldots \partial_{m_{|I|}}^{-1} z_{l_1}^{\mathcal{B}^T_{I,I\cup\{l_1\}}} \cdot \ldots \cdot z_{l_{n-|I|}}^{\mathcal{B}^T_{I,I\cup\{l_{n-|I|}\}}} \cdot z_{m_1}^{\mathcal{B}^T_{I,I\setminus\{m_1\}}-\mathrm{Id}} \cdot \ldots \cdot z_{m_{|I|}}^{\mathcal{B}^T_{I,I\setminus\{m_{|I|}\}}-\mathrm{Id}} \cdot \alpha^T \in (\Lambda_{I,0})^n.$$

The eigenvalues of $\mathcal{B}^T_{I,I\setminus\{m\}} - \mathrm{Id}$ are contained in Σ for $m \in I$, as $\mathcal{V}$ is an object in $\mathcal{Q}ui_n^{\Sigma_1}$. Now, Lemma 4.4 yields $\alpha \equiv 0$ and η_I is injective.

As $\dim_{\mathbb{C}} \mathrm{Hom}_{\mathscr{D}_0}((E\mathcal{V})_0, \mathcal{O}_{I,0}) \leq n_I$ by part (iv), we immediately obtain the bijectivity of η_I as claimed.

(vi) Let us summarize: In part (i) we saw that

$$\begin{aligned}
&\mathrm{Hom}_{\mathscr{D}_0}((E\mathcal{V})_0, \mathcal{O}_{I,0}) \cong \\
&\cong \Bigg\{ \bigoplus_{J \in \mathcal{P}(\{1,\dots,n\})} \phi^J \in \bigoplus_{J \in \mathcal{P}(\{1,\dots,n\})} \mathrm{Hom}_{\mathbb{C}}(V_J, \mathcal{O}_{I,0}) \;\Bigg|\; \\
&\qquad \partial_j \cdot \phi^J(v_J) - \phi^{J\cup\{j\}}(B_{J\cup\{j\},J}(v_J)) = 0, \\
&\qquad z_j \cdot \phi^{J\cup\{j\}}(v_{J\cup\{j\}}) - \phi^J(B_{J,J\cup\{j\}}(v_{J\cup\{j\}})) = 0 \\
&\qquad \text{for } J \neq \{1,\dots,n\},\ j \in \{1,\dots,n\} \setminus J,\ v_J \in V_J,\ v_{J\cup\{j\}} \in V_{J\cup\{j\}} \Bigg\}.
\end{aligned}$$

Part (ii) shows that in fact all ϕ^J can be uniquely recovered from ϕ^I using (ALG) and we have

$$\begin{aligned}
&\mathrm{Hom}_{\mathscr{D}_0}((E\mathcal{V})_0, \mathcal{O}_{I,0}) \cong \\
&\cong \Big\{ \phi^I \in \mathrm{Hom}_{\mathbb{C}}(V_I, \mathcal{O}_{I,0}) \;\Big|\; z_l\partial_l \cdot \phi^I(v_I) - \phi^I(B_{I,I\cup\{l\}}B_{I\cup\{l\},I}(v_I)) = 0, \\
&\qquad z_m\partial_m \cdot \phi^I(v_I) - \phi^I((B_{I,I\setminus\{m\}}B_{I\setminus\{m\},I} - \mathrm{Id})(v_I)) = 0 \qquad (1) \\
&\qquad \text{for } l \in \{1,\dots,n\} \setminus I,\, m \in I,\, v_I \in V_I \Big\}.
\end{aligned}$$

In part (v) we gave a canonical morphism η_I from V_I^* into (1). We showed that η_I is injective. Finally, using part (iv) we obtained the bijectivity.
All in all, this gives us the canonical isomorphism

$$\mathfrak{a}\colon V_I^* \xrightarrow{\cong} \mathrm{Hom}_{\mathscr{D}_0}((E\mathcal{V})_0, \mathcal{O}_{I,0})\,. \qquad \square$$

So, we saw that the vector space $\mathrm{Hom}_{\mathscr{D}_0}((E\mathcal{V})_0, \mathcal{O}_{I,0})$ can be easily recovered from $\mathcal{V}$ as it is just the dual of the vector space V_I.

The following statement on the matrix polynomial will be used in the proof of Proposition 4.8. We state it here as it is rather independent of our quiver $\mathscr{D}$-modules.

Corollary 4.7.
Let A denote a square matrix with entries in $\mathbb{C}$ and let $i \in \{1,\dots,n\}$. We fix a branch of the logarithm defined on $\mathbb{C}_i \setminus \mathbb{R}_0^+$ and let $z_i^A = \exp(A \cdot \ln(z_i))$ as before. We define

$$\varphi_A(z_i) := \sum_{k=0}^{\infty} \frac{A^k}{(k+1)!} \cdot \ln(z_i)^{k+1} \quad \textit{and} \quad \psi(A) := \sum_{k=1}^{\infty} \frac{(2\pi i)^k}{k!} A^{k-1}\,.$$

Then

$$M_i\varphi_A(z_i) - \varphi_A(z_i) = \psi(A) \cdot z_i^A$$

where $M_i\varphi_A(z_i)$ is the function after analytic continuation of $\varphi_A(z_i)$ around $z_i = 0$.

Proof:
We clearly have

$$M_i\varphi_A(z_i) = \sum_{k=0}^{\infty} \frac{A^k}{(k+1)!} \cdot (\ln(z_i) + 2\pi i)^{k+1}.$$

Let J denote the Jordan normal form of A, and let B denote the invertible matrix such that $A = BJB^{-1}$. Then

$$\varphi_A(z_i) = B \cdot \varphi_J(z_i) \cdot B^{-1}, \quad M_i\varphi_A(z_i) = B \cdot M_i\varphi_J(z_i) \cdot B^{-1} \;\text{ and }\; \psi(A) \cdot z_i^A = B \cdot \psi(J) \cdot z_i^J \cdot B^{-1}.$$

So, we may assume without loss of generality that A is given in Jordan normal form. Furthermore, we see that the left and right hand side of our claimed equation respect the blocks of the Jordan normal form and we may assume that A is given as a single Jordan block.

Obviously, it yields

$$A \cdot (M_i \varphi_A(z_i) - \varphi_A(z_i)) = (M_i \varphi_A(z_i) - \varphi_A(z_i)) \cdot A \quad \text{and} \quad A \cdot \psi(A) \cdot z_i^A = \psi(A) \cdot z_i^A \cdot A\,.$$

But we also have the equality

$$\begin{aligned} A \cdot (M_i \varphi_A(z_i) - \varphi_A(z_i)) &= \exp(A \ln(z_i) + 2\pi i A) - \exp(A \ln(z_i)) = \\ &= (\exp(2\pi i A) - \mathrm{Id}) \cdot z_i^A = A \cdot \psi(A) \cdot z_i^A \end{aligned}$$

which gives us

$$(M_i \varphi_A(z_i) - \varphi_A(z_i)) \cdot A = \psi(A) \cdot z_i^A \cdot A\,.$$

If A is invertible, our claim follows immediately. So assume that A is not invertible, i. e. A is the Jordan block with 0 on the diagonal and 1 on the first upper off-diagonal. The above two equations show us however that $M_i \varphi_A(z_i) - \varphi_A(z_i)$ and $\psi(A) \cdot z_i^A$ coincide up to a possible difference in the entry in the upper-left corner. The entry of $M_i \varphi_A(z_i) - \varphi_A(z_i)$ in the upper-left corner is $\ln(z_i) + 2\pi i - \ln(z_i) = 2\pi i$. The first column of z_i^A is $(1, 0, \ldots, 0)^T$ and the entry in the upper-left corner of $\psi(A)$ is $2\pi i$. This shows us that the entry in the upper-left corner of $\psi(A) \cdot z_i^A$ is $2\pi i$ as well, which gives us the claimed equality $M_i \varphi_A(z_i) - \varphi_A(z_i) = \psi(A) \cdot z_i^A$. □

In the following proposition we are going to prove that the quiver representation one obtains after applying $\mathscr{A}$ to a quiver $\mathscr{D}$-module is determined in a very simple manner by the starting quiver representation. To do so, we use the canonical isomorphism $\mathfrak{a}$ from Proposition 4.6 and "extend" it to the whole quiver representation. As before we use the category $\mathcal{Q}ui_n^{\Sigma_1}$. We will see that the morphisms which are part of the terminating quiver representation are similar to those in dimension 1 from Chapter 3.

Proposition 4.8.
Let $\mathcal{V}_n = (V_I, B_{I\cup\{i\},I}, B_{I,I\cup\{i\}})$ denote an object of the quiver representation category $\mathcal{Q}ui_n^{\Sigma_1}$ and let $E\mathcal{V}_n$ denote the corresponding quiver $\mathscr{D}$-module. The image of $E\mathcal{V}_n$ under the functor $\mathscr{A}$ is canonically isomorphic to

$$V_I^* \underset{w_{I,i}}{\overset{u_{I,i}}{\rightleftarrows}} V_{I\cup\{i\}}^*$$

where the maps $u_{I,i}$ and $w_{I,i}$ are given by

$$u_{I,i} = B_{I,I\cup\{i\}}^* \quad \text{and} \quad w_{I,i} = B_{I\cup\{i\},I}^* \circ \sum_{k=1}^{\infty} \frac{(2\pi i)^k}{k!} (B_{I,I\cup\{i\}}^* \circ B_{I\cup\{i\},I}^*)^{k-1}\,.$$

Proof:
Recall that $n_I = \dim_{\mathbb{C}} V_I$, and $\mathcal{B}_{K,L} = B_{K,L} \circ B_{L,K}$ if $K, L \in \mathcal{P}(\{1, \ldots, n\})$ are adjacent. The image of $(V_I, B_{I\cup\{i\},I}, B_{I,I\cup\{i\}})$ under the functor $\mathcal{A} \circ E$ is given by

$$(\mathrm{Hom}_{\mathscr{D}_0}((E\mathcal{V}_n)_0, \mathcal{O}_{I,0}), \mathrm{can}_{I,i}, \mathrm{var}_{I,i}).$$

We note that we will partly omit set braces for singletons to improve the readability. First, we reperform the main steps of the proof of Proposition 4.6. Then we compute the canonical map and the variation.

(i) First of all, note that the natural isomorphisms we gave for $\mathrm{Hom}_{\mathscr{D}_0}((E\mathcal{V}_n)_0, \mathcal{O}_{I,0})$ in part (i) of the proof of Proposition 4.6 are compatible with the canonical map and the variation. Therefore, they extend to the entire object $(\mathrm{Hom}_{\mathscr{D}_0}((E\mathcal{V}_n)_0, \mathcal{O}_{I,0}), \mathrm{can}_{I,i}, \mathrm{var}_{I,i})$.
We obtain:

$$\mathrm{Hom}_{\mathscr{D}_0}((E\mathcal{V}_n)_0, \mathcal{O}_{I,0}) \underset{\mathrm{var}_{I,i}}{\overset{\mathrm{can}_{I,i}}{\rightleftarrows}} \mathrm{Hom}_{\mathscr{D}_0}((E\mathcal{V}_n)_0, \mathcal{O}_{I\cup\{i\},0}) \cong$$

$$\left\{ \bigoplus_J \phi_I^J \in \bigoplus_J \mathrm{Hom}_{\mathbb{C}}(V_J, \mathcal{O}_{I,0}) \,\middle|\, \begin{array}{l} \partial_j \cdot \phi_I^J(v_J) - \phi_I^{J\cup\{j\}}(B_{J\cup\{j\},J}(v_J)) = 0, \\ z_j \cdot \phi_I^{J\cup\{j\}}(v_{J\cup\{j\}}) - \phi_I^J(B_{J,J\cup\{j\}}(v_{J\cup\{j\}})) = 0 \\ \text{for } J \neq \{1,\dots,n\},\, j \in \{1,\dots,n\}\setminus J,\, v_J \in V_J,\, v_{J\cup\{j\}} \in V_{J\cup\{j\}} \end{array} \right\}$$

$$\bigoplus_J \mathrm{can}_{I,i}^J \downarrow \;\uparrow \bigoplus_J \mathrm{var}_{I,i}^J$$

$$\left\{ \bigoplus_J \phi_{I\cup\{i\}}^J \in \bigoplus_J \mathrm{Hom}_{\mathbb{C}}(V_J, \mathcal{O}_{I\cup\{i\},0}) \,\middle|\, \begin{array}{l} \partial_j \cdot \phi_{I\cup\{i\}}^J(v_J) - \phi_{I\cup\{i\}}^{J\cup\{j\}}(B_{J\cup\{j\},J}(v_J)) = 0, \\ z_j \cdot \phi_{I\cup\{i\}}^{J\cup\{j\}}(v_{J\cup\{j\}}) - \phi_{I\cup\{i\}}^J(B_{J,J\cup\{j\}}(v_{J\cup\{j\}})) = 0 \\ \text{for } J \neq \{1,\dots,n\},\, j \in \{1,\dots,n\}\setminus J,\, v_J \in V_J,\, v_{J\cup\{j\}} \in V_{J\cup\{j\}} \end{array} \right\}$$

(ii) Let us fix $I \in \mathcal{P}(\{1,\dots,n\})\setminus\{1,\dots,n\}$, $i \in \{1,\dots,n\}\setminus I$ temporarily. We saw in part (ii) of the proof of Proposition 4.6 that all ϕ_I^J and $\phi_{I\cup\{i\}}^J$ can be uniquely recovered from ϕ_I^I and $\phi_{I\cup\{i\}}^{I\cup\{i\}}$, respectively, using (ALG). This allows us to consider only the behaviour of the following two pairs

$$\begin{pmatrix} \phi_I^I \\ \phi_I^{I\cup\{i\}} \end{pmatrix} \begin{matrix} \leftrightarrow \\ \leftrightarrow \end{matrix} \begin{pmatrix} \phi_{I\cup\{i\}}^I \\ \phi_{I\cup\{i\}}^{I\cup\{i\}} \end{pmatrix}$$

under the canonical map and the variation for the following reason:
Both pairs arise of the same system of equations where $J \in \mathcal{P}(\{1,\dots,n\})\setminus\{1,\dots,n\}$, $j \in \{1,\dots,n\}\setminus J$:

$$\begin{aligned} \partial_j \cdot \phi^J - \phi^{J\cup\{j\}} \circ B_{J\cup\{j\},J} &= 0 \\ z_j \cdot \phi^{J\cup\{j\}} - \phi^J \circ B_{J,J\cup\{j\}} &= 0 \end{aligned} \qquad (\star)$$

The main difference is the target space of the maps: ϕ_I^I and $\phi_I^{I\cup i}$ map to $\mathcal{O}_{I,0}$, whereas $\phi_{I\cup i}^I$ and $\phi_{I\cup i}^{I\cup i}$ map to $\mathcal{O}_{I\cup i,0}$. By Lemma 4.3, z_j acts bijective on $\mathcal{O}_{I,0}$ and $\mathcal{O}_{I\cup i,0}$ if and only if $j \notin I\cup\{i\}$, and ∂_k acts bijective on $\mathcal{O}_{I,0}$ and $\mathcal{O}_{I\cup i,0}$ if and only if $k \in I$. This allows us to adapt the algorithm (ALG) from the proof of Proposition 4.6 in the following way: We only use equations of ($\star$) which involve z_j for $j \notin I\cup\{i\}$ or ∂_k for $k \in I$ to express every ϕ_I^K uniquely in terms of ϕ_I^I or in terms of $\phi_I^{I\cup i}$.

We obtain for $\blacksquare = I$ or $I \cup \{i\}$ and any $K \in \mathcal{P}(\{1,\dots,n\})$ that

$$\phi_{\blacksquare}^{K} = \begin{cases} (z_{K\setminus I})^{-1} \cdot (\partial_{I\setminus K})^{-1} \cdot (\phi_{\blacksquare}^{I} \circ A^{I}) & \text{if } i \notin K, \\ (z_{K\setminus\{I\cup i\}})^{-1} \cdot (\partial_{\{I\cup i\}\setminus K})^{-1} \cdot (\phi_{\blacksquare}^{I\cup i} \circ A^{I\cup i}) & \text{if } i \in K. \end{cases}$$

$A^{I} \in \mathrm{Hom}_{\mathbb{C}}(V_K, V_I)$ and $A^{I\cup i} \in \mathrm{Hom}_{\mathbb{C}}(V_K, V_{I\cup i})$ are compositions of linear maps $B_{\bullet,\bullet}$ and they do not depend on $\blacksquare$. We abbreviated for $J \in \mathcal{P}(\{1,\dots,n\})$, $J = \{j_1,\dots,j_{|J|}\}$

$$(z_J)^{-1} := z_{j_1}^{-1} \cdot \ldots \cdot z_{j_{|J|}}^{-1} \quad \text{and} \quad (\partial_J)^{-1} := \partial_{j_1}^{-1} \cdot \ldots \cdot \partial_{j_{|J|}}^{-1}.$$

The uniqueness of the expression for $\phi_{\bullet}^{K}$ follows similarly to the argumentation for (ALG). We see that ϕ_I^K and $\phi_{I\cup i}^K$ are build up from ϕ_I^I and $\phi_{I\cup i}^I$ for $i \notin K$ (from $\phi_I^{I\cup i}$ and $\phi_{I\cup i}^{I\cup i}$ for $i \notin K$) in a completely identical manner. This shows us that it suffices to examine the behaviour of the two pairs

$$\begin{pmatrix} \phi_I^I \\ \phi_I^{I\cup i} \end{pmatrix} \begin{matrix} \leftrightarrow \\ \leftrightarrow \end{matrix} \begin{pmatrix} \phi_{I\cup i}^I \\ \phi_{I\cup i}^{I\cup i} \end{pmatrix}$$

under the canonical map and the variation. For any $K \in \mathcal{P}(\{1,\dots,n\})$, $\phi_I^K \leftrightarrow \phi_{I\cup i}^K$ will follow their behaviour under these two maps.

(iii) In part (v) of the proof of Proposition 4.6 we have seen that ϕ_I^I and $\phi_{I\cup i}^{I\cup i}$ can be uniquely identified with elements of V_I^* and $V_{I\cup i}^*$, respectively, using the canonical isomorphisms η_I and $\eta_{I\cup i}$, respectively. After a choice of basis of V_I and $V_{I\cup i}$, we may write (using (ALG))

$$\phi_I^I = \eta_I(\alpha_I) = \alpha_I \cdot \mathscr{F}_I \qquad \phi_{I\cup i}^I = \partial_i^{-1} \cdot \phi_{I\cup i}^{I\cup i} \cdot B_{I\cup i,I}$$
$$\phi_I^{I\cup i} = z_i^{-1} \cdot \phi_I^I \cdot B_{I,I\cup i} \qquad \phi_{I\cup i}^{I\cup i} = \eta_{I\cup i}(\alpha_{I\cup i}) = \alpha_{I\cup i} \cdot \mathscr{F}_{I\cup i}$$

for some $\alpha_I \in V_I^*$, $\alpha_{I\cup i} \in V_{I\cup i}^*$ and

$$\mathscr{F}_I = z_i^{\mathcal{B}_{I,I\cup i}} \cdot z_{l_2}^{\mathcal{B}_{I,I\cup l_2}} \cdot \ldots \cdot z_{l_{n-|I|}}^{\mathcal{B}_{I,I\cup l_{n-|I|}}} \cdot z_{m_1}^{\mathcal{B}_{I,I\setminus m_1}-\mathrm{Id}} \cdot \ldots \cdot z_{m_{|I|}}^{\mathcal{B}_{I,I\setminus m_{|I|}}-\mathrm{Id}}$$
$$\mathscr{F}_{I\cup i} = z_{l_2}^{\mathcal{B}_{I\cup i,I\cup\{i,l_2\}}} \cdot \ldots \cdot z_{l_{n-|I|}}^{\mathcal{B}_{I\cup i,I\cup\{i,l_{n-|I|}\}}} \cdot z_i^{\mathcal{B}_{I\cup i,I}-\mathrm{Id}} \cdot z_{m_1}^{\mathcal{B}_{I\cup i,\{I\cup i\}\setminus m_1}-\mathrm{Id}} \cdot \ldots \cdot z_{m_{|I|}}^{\mathcal{B}_{I\cup i,\{I\cup i\}\setminus m_{|I|}}-\mathrm{Id}}$$

where $\{i,l_2,\dots,l_{n-|I|}\} = \{1,\dots,n\} \setminus I$, $\{m_1,\dots,m_{|I|}\} = I$. $\mathscr{F}_I$ and $\mathscr{F}_{I\cup i}$ are matrices of size $n_I \times n_I$ and $n_{I\cup i} \times n_{I\cup i}$, respectively, with entries in $\mathcal{O}_{I,0}$ and $\mathcal{O}_{I\cup i,0}$, respectively. This description of $\phi_{I\cup i}^I$ and $\phi_I^{I\cup i}$ is independent of the choice of the basis just as the one of ϕ_I^I and $\phi_{I\cup i}^{I\cup i}$.

Let us give some identities which will be helpful for our further computations: We already stated that for $K \in \mathcal{P}(\{1,\dots,n\})$

$$\mathcal{B}_{K,L_1} \cdot \mathcal{B}_{K,L_2} = \mathcal{B}_{K,L_2} \cdot \mathcal{B}_{K,L_1}$$

if $L_1, L_2 \in \mathcal{P}(\{1,\dots,n\})$ are adjacent to K. And we have for $i, l \notin I$, $l \neq i$, $m \in I$:

$$B_{I,I\cup i} \cdot \mathcal{B}_{I\cup i,\{I\cup i\}\setminus m} = \mathcal{B}_{I,I\setminus m} \cdot B_{I,I\cup i} \qquad B_{I\cup i,I} \cdot \mathcal{B}_{I,I\cup l} = \mathcal{B}_{I\cup i,I\cup\{i,l\}} \cdot B_{I\cup i,I}$$
$$B_{I,I\cup i} \cdot \mathcal{B}_{I\cup i,I\cup\{i,l\}} = \mathcal{B}_{I,I\cup l} \cdot B_{I,I\cup i} \qquad B_{I\cup i,I} \cdot \mathcal{B}_{I,I\setminus m} = \mathcal{B}_{I\cup i,\{I\cup i\}\setminus m} \cdot B_{I\cup i,I}$$

(iv) We claimed that the canonical map from $\begin{pmatrix}\phi_I^I\\ \phi_I^{I\cup i}\end{pmatrix}$ to $\begin{pmatrix}\phi_{I\cup i}^I\\ \phi_{I\cup i}^{I\cup i}\end{pmatrix}$ is given by $B^*_{I,I\cup i}$. This means we have to check that the assignment

$$\alpha_I \mapsto \alpha_{I\cup i} := \alpha_I \cdot B_{I,I\cup i}$$

describes the canonical map. This follows by direct computations.

We have

$$\begin{aligned}
&\eta_{I\cup i}(\alpha_I \cdot B_{I,I\cup i}) = \alpha_I \cdot B_{I,I\cup i} \cdot \mathscr{F}_{I\cup i} = \\
&= \alpha_I \cdot B_{I,I\cup i} \cdot z_{l_2}^{\mathcal{B}_{I\cup i,I\cup\{i,l_2\}}} \cdot \ldots \cdot z_{l_{n-|I|}}^{\mathcal{B}_{I\cup i,I\cup\{i,l_{n-|I|}\}}} \cdot z_i^{\mathcal{B}_{I\cup i,I}-\mathrm{Id}} \cdot \\
&\cdot z_{m_1}^{\mathcal{B}_{I\cup i,\{I\cup i\}\setminus m_1}-\mathrm{Id}} \cdot \ldots \cdot z_{m_{|I|}}^{\mathcal{B}_{I\cup i,\{I\cup i\}\setminus m_{|I|}}-\mathrm{Id}} = \\
&= \alpha_I \cdot z_{l_2}^{\mathcal{B}_{I,I\cup l_2}} \cdot \ldots \cdot z_{l_{n-|I|}}^{\mathcal{B}_{I,I\cup l_{n-|I|}}} \cdot z_i^{-1} \cdot z_i^{\mathcal{B}_{I,I\cup i}} \cdot B_{I,I\cup i} \cdot \\
&\cdot z_{m_1}^{\mathcal{B}_{I\cup i,\{I\cup i\}\setminus m_1}-\mathrm{Id}} \cdot \ldots \cdot z_{m_{|I|}}^{\mathcal{B}_{I\cup i,\{I\cup i\}\setminus m_{|I|}}-\mathrm{Id}} = \\
&= z_i^{-1} \cdot \alpha_I \cdot z_i^{\mathcal{B}_{I,I\cup i}} \cdot z_{l_2}^{\mathcal{B}_{I,I\cup l_2}} \cdot \ldots \cdot z_{l_{n-|I|}}^{\mathcal{B}_{I,I\cup l_{n-|I|}}} \cdot z_{m_1}^{\mathcal{B}_{I,I\setminus m_1}-\mathrm{Id}} \cdot \ldots \cdot z_{m_{|I|}}^{\mathcal{B}_{I,I\setminus m_{|I|}}-\mathrm{Id}} \cdot B_{I,I\cup i} = \\
&= z_i^{-1} \cdot \alpha_I \cdot \mathscr{F}_I \cdot B_{I,I\cup i} = z_i^{-1} \cdot \phi_I^I \cdot B_{I,I\cup i} = \phi_I^{I\cup i}
\end{aligned}$$

and therefore

$$\begin{aligned}
&\eth_i^{-1} \cdot \eta_{I\cup i}(\alpha_I \cdot B_{I,I\cup i}) \cdot B_{I\cup i,I} = \eth_i^{-1} z_i^{-1} \cdot \alpha_I \cdot \mathscr{F}_I \cdot B_{I,I\cup i} \cdot B_{I\cup i,I} = \\
&= \alpha_I \cdot \eth_i^{-1} z_i^{-1} \cdot B_{I,I\cup i} \cdot z_i^{\mathcal{B}_{I,I\cup i}} \cdot z_{l_2}^{\mathcal{B}_{I,I\cup l_2}} \cdot \ldots \cdot z_{l_{n-|I|}}^{\mathcal{B}_{I,I\cup l_{n-|I|}}} \cdot z_{m_1}^{\mathcal{B}_{I,I\setminus m_1}-\mathrm{Id}} \cdot \ldots \cdot z_{m_{|I|}}^{\mathcal{B}_{I,I\setminus m_{|I|}}-\mathrm{Id}} = \\
&= \alpha_I \cdot z_i^{\mathcal{B}_{I,I\cup i}} \cdot z_{l_2}^{\mathcal{B}_{I,I\cup l_2}} \cdot \ldots \cdot z_{l_{n-|I|}}^{\mathcal{B}_{I,I\cup l_{n-|I|}}} \cdot z_{m_1}^{\mathcal{B}_{I,I\setminus m_1}-\mathrm{Id}} \cdot \ldots \cdot z_{m_{|I|}}^{\mathcal{B}_{I,I\setminus m_{|I|}}-\mathrm{Id}} = \alpha_I \cdot \mathscr{F}_I = \phi_I^I
\end{aligned}$$

as asserted. Just as in step (v) of the proof of Proposition 4.6, it is easily checked that the description of the canonical map does not depend on the choice of basis.

(v) We are left with the computation of the variation of $\begin{pmatrix}\phi_{I\cup i}^I\\ \phi_{I\cup i}^{I\cup i}\end{pmatrix}$, i.e.

$$M_i\phi_{I\cup i}^I - \phi_{I\cup i}^I \quad \text{and} \quad M_i\phi_{I\cup i}^{I\cup i} - \phi_{I\cup i}^{I\cup i}.$$

In particular, we need to check that the assignment

$$\alpha_{I\cup i} \mapsto \alpha_I := \alpha_{I\cup i} \cdot \left(\sum_{k=1}^{\infty} \frac{(2\pi i)^k}{k!} (B_{I\cup i,I} B_{I,I\cup i})^{k-1}\right) \cdot B_{I\cup i,I} = \alpha_{I\cup i} \cdot \underbrace{\psi(\mathcal{B}_{I\cup i,I}) \cdot B_{I\cup i,I}}_{=:\Theta_{I,i}}$$

describes the variation. For $\phi_{I\cup i}^{I\cup i}$ the correctness follows by direct computation:

$$\begin{aligned}
&M_i\phi_{I\cup i}^{I\cup i} - \phi_{I\cup i}^{I\cup i} = \\
&= \alpha_{I\cup i} z_{l_2}^{\mathcal{B}_{I\cup i,I\cup\{i,l_2\}}} \ldots z_{l_{n-|I|}}^{\mathcal{B}_{I\cup i,I\cup\{i,l_{n-|I|}\}}} z_i^{\mathcal{B}_{I\cup i,I}-\mathrm{Id}} e^{2\pi i \mathcal{B}_{I\cup i,I}} z_{m_1}^{\mathcal{B}_{I\cup i,\{I\cup i\}\setminus m_1}-\mathrm{Id}} \ldots z_{m_{|I|}}^{\mathcal{B}_{I\cup i,\{I\cup i\}\setminus m_{|I|}}-\mathrm{Id}} + \\
&- \alpha_{I\cup i} z_{l_2}^{\mathcal{B}_{I\cup i,I\cup\{i,l_2\}}} \ldots z_{l_{n-|I|}}^{\mathcal{B}_{I\cup i,I\cup\{i,l_{n-|I|}\}}} z_i^{\mathcal{B}_{I\cup i,I}-\mathrm{Id}} z_{m_1}^{\mathcal{B}_{I\cup i,\{I\cup i\}\setminus m_1}-\mathrm{Id}} \ldots z_{m_{|I|}}^{\mathcal{B}_{I\cup i,\{I\cup i\}\setminus m_{|I|}}-\mathrm{Id}} =
\end{aligned}$$

$$
\begin{aligned}
&= \alpha_{I\cup i} \cdot (e^{2\pi i \mathcal{B}_{I\cup i,I}} - \mathrm{Id}) \cdot \mathscr{F}_{I\cup i} = \\
&= \alpha_{I\cup i} \cdot \sum_{k=1}^{\infty} \frac{(2\pi i)^k}{k!} (B_{I\cup i,I} B_{I,I\cup i})^{k-1} \cdot B_{I\cup i,I} \cdot B_{I,I\cup i} \cdot \mathscr{F}_{I\cup i} = \\
&= \alpha_{I\cup i} \cdot \sum_{k=1}^{\infty} \frac{(2\pi i)^k}{k!} (B_{I\cup i,I} B_{I,I\cup i})^{k-1} \cdot B_{I\cup i,I} \cdot z_i^{-1} \cdot \mathscr{F}_I \cdot B_{I,I\cup i} = \\
&= z_i^{-1} \cdot \alpha_{I\cup i} \cdot \Theta_{I,i} \cdot \mathscr{F}_I \cdot B_{I,I\cup i} = z_i^{-1} \cdot \eta_I(\alpha_{I\cup i} \cdot \Theta_{I,i}) \cdot B_{I,I\cup i}
\end{aligned}
$$

Now, let us compute $M_i \phi^I_{I\cup i} - \phi^I_{I\cup i}$: We use the identity

$$
\begin{aligned}
&\mathscr{F}_{I\cup i} \cdot B_{I\cup i,I} = \\
&= z_{l_2}^{\mathcal{B}_{I\cup i, I\cup\{i,l_2\}}} \dots z_{l_{n-|I|}}^{\mathcal{B}_{I\cup i, I\cup\{i,l_{n-|I|}\}}} z_i^{\mathcal{B}_{I\cup i,I}-\mathrm{Id}} z_{m_1}^{\mathcal{B}_{I\cup i,\{I\cup i\}\setminus m_1}-\mathrm{Id}} \dots z_{m_{|I|}}^{\mathcal{B}_{I\cup i,\{I\cup i\}\setminus m_{|I|}}-\mathrm{Id}} B_{I\cup i,I} = \\
&= z_i^{-1} \cdot B_{I\cup i,I} \cdot z_{l_2}^{\mathcal{B}_{I,I\cup\{l_2\}}} \cdot \dots \cdot z_{l_{n-|I|}}^{\mathcal{B}_{I,I\cup\{l_{n-|I|}\}}} z_i^{\mathcal{B}_{I,I\cup i}} \cdot z_{m_1}^{\mathcal{B}_{I,I\setminus m_1}-\mathrm{Id}} \cdot \dots \cdot z_{m_{|I|}}^{\mathcal{B}_{I,I\setminus m_{|I|}}-\mathrm{Id}} = \\
&= z_i^{-1} \cdot B_{I\cup i,I} \cdot \mathscr{F}_I
\end{aligned}
$$

to rearrange $\phi^I_{I\cup i}$. We obtain

$$
\begin{aligned}
\phi^I_{I\cup i} &= \partial_i^{-1} \cdot \alpha_{I\cup i} \cdot \mathscr{F}_{I\cup i} \cdot B_{I\cup i,I} = \alpha_{I\cup i} \cdot \partial_i^{-1} z_i^{-1} \cdot B_{I\cup i,I} \cdot \mathscr{F}_I = \\
&= \alpha_{I\cup i} B_{I\cup i,I} \cdot \left(\partial_i^{-1} z_i^{-1} z_i^{\mathcal{B}_{I,I\cup i}} \right) \cdot z_{l_2}^{\mathcal{B}_{I,I\cup l_2}} \cdot \dots \cdot z_{l_{n-|I|}}^{\mathcal{B}_{I,I\cup l_{n-|I|}}} \cdot z_{m_1}^{\mathcal{B}_{I,I\setminus m_1}-\mathrm{Id}} \cdot \dots \cdot z_{m_{|I|}}^{\mathcal{B}_{I,I\setminus m_{|I|}}-\mathrm{Id}} .
\end{aligned}
$$

Apart from that, we have

$$
\partial_i^{-1} z_i^{-1} z_i^{\mathcal{B}_{I,I\cup i}} = \partial_i^{-1} \sum_{k=0}^{\infty} \frac{\mathcal{B}^k_{I,I\cup i}}{k!} \frac{\ln(z_i)^k}{z_i} = \sum_{k=0}^{\infty} \frac{\mathcal{B}^k_{I,I\cup i}}{k!} \frac{\ln(z_i)^{k+1}}{k+1} =: \varphi_{\mathcal{B}_{I,I\cup i}}(z_i) .
$$

Corollary 4.7 yields

$$
M_i \varphi_{\mathcal{B}_{I,I\cup i}}(z_i) - \varphi_{\mathcal{B}_{I,I\cup i}}(z_i) = \left(\sum_{k=1}^{\infty} \frac{(2\pi i)^k}{k!} \cdot \mathcal{B}^{k-1}_{I,I\cup i} \right) \cdot z_i^{\mathcal{B}_{I,I\cup i}} .
$$

This gives us

$$
\begin{aligned}
&M_i \phi^I_{I\cup i} - \phi^I_{I\cup i} = \\
&= \alpha_{I\cup i} B_{I\cup i,I} \left(\sum_{k=1}^{\infty} \frac{(2\pi i)^k}{k!} \cdot \mathcal{B}^{k-1}_{I,I\cup i} \right) z_i^{\mathcal{B}_{I,I\cup i}} z_{l_2}^{\mathcal{B}_{I,I\cup l_2}} \dots z_{l_{n-|I|}}^{\mathcal{B}_{I,I\cup l_{n-|I|}}} z_{m_1}^{\mathcal{B}_{I,I\setminus m_1}-\mathrm{Id}} \dots z_{m_{|I|}}^{\mathcal{B}_{I,I\setminus m_{|I|}}-\mathrm{Id}} = \\
&= \alpha_{I\cup i} \left(\sum_{k=1}^{\infty} \frac{(2\pi i)^k}{k!} \cdot \mathcal{B}^{k-1}_{I\cup i,I} \right) B_{I\cup i,I} \, z_i^{\mathcal{B}_{I,I\cup i}} z_{l_2}^{\mathcal{B}_{I,I\cup l_2}} \dots z_{l_{n-|I|}}^{\mathcal{B}_{I,I\cup l_{n-|I|}}} z_{m_1}^{\mathcal{B}_{I,I\setminus m_1}-\mathrm{Id}} \dots z_{m_{|I|}}^{\mathcal{B}_{I,I\setminus m_{|I|}}-\mathrm{Id}} = \\
&= \alpha_{I\cup i} \cdot \Theta_{I,i} \cdot \mathscr{F}_I = \eta_I(\alpha_{I\cup i} \cdot \Theta_{I,i})
\end{aligned}
$$

as claimed. Once more, note that these computations are independent of the choice of basis.

All in all, we obtain the claimed isomorphism. □

Now, we have collected all the important pieces for the proof of our Main Theorem 4.5:

Proof of Theorem 4.5:
Proposition 4.8 yields an isomorphism for every object. So, let us examine if this family of isomorphisms is natural:
First of all, we note that the isomorphisms we gave in part (i) of the proof of Proposition 4.8 are natural isomorphisms. So let $\mathcal{V} = (V_J, B_{J\cup\{j\},J}, B_{J,J\cup\{j\}})$ and $\tilde{\mathcal{V}} = (\tilde{V}_J, \tilde{B}_{J\cup\{j\},J}, \tilde{B}_{J,J\cup\{j\}})$ denote two objects in $\mathcal{Q}ui_n^{\Sigma_1}$ and let $\tau = (h_J)$ denote a morphism from $\mathcal{V}$ to $\tilde{\mathcal{V}}$. We need to check that the diagram

$$\begin{array}{ccc}
\{\oplus_J\phi_I^J \in \oplus_J \mathrm{Hom}_{\mathbb{C}}(V_J, \mathcal{O}_{I,0}) \mid \ldots\} & \xleftarrow{(\mathrm{Hom}_{\mathbb{C}}(\tau, \mathcal{O}_{I,0}))} & \{\oplus_J\tilde{\phi}_I^J \in \oplus_J \mathrm{Hom}_{\mathbb{C}}(\widetilde{V}_J, \mathcal{O}_{I,0}) \mid \ldots\} \\
\oplus_J \mathrm{can}_{I,i}^J \downarrow \uparrow \oplus_J \mathrm{var}_{I,i}^J & & \oplus_J\widetilde{\mathrm{can}}_{I,i}^J \downarrow \uparrow \oplus_J\widetilde{\mathrm{var}}_{I,i}^J \\
\{\oplus_J\phi_{I\cup\{i\}}^J \in \oplus_J \mathrm{Hom}_{\mathbb{C}}(V_J, \mathcal{O}_{I\cup\{i\},0}) \mid \ldots\} & & \{\oplus_J\tilde{\phi}_{I\cup\{i\}}^J \in \oplus_J \mathrm{Hom}_{\mathbb{C}}(\widetilde{V}_J, \mathcal{O}_{I\cup\{i\},0}) \mid \ldots\} \\
\uparrow & & \uparrow \\
\mathrm{Hom}_{\mathbb{C}}(V_I, \mathbb{C}) & \xleftarrow{(h_I^*)} & \mathrm{Hom}_{\mathbb{C}}(\widetilde{V}_I, \mathbb{C}) \\
B_{I,I\cup\{i\}}^* \downarrow \uparrow \psi(B_{I,I\cup\{i\}}^*) \circ B_{I\cup\{i\},I}^* & & \tilde{B}_{I,I\cup\{i\}}^* \downarrow \uparrow \psi(\tilde{B}_{I,I\cup\{i\}}^*) \circ \tilde{B}_{I\cup\{i\},I}^* \\
\mathrm{Hom}_{\mathbb{C}}(V_{I\cup\{i\}}, \mathbb{C}) & & \mathrm{Hom}_{\mathbb{C}}(\widetilde{V}_{I\cup\{i\}}, \mathbb{C})
\end{array}$$

commutes. The properties indicated by "..." in the upper row can be found in part (i) of the proof of Proposition 4.8. We omitted them here to keep the diagram legible. The morphisms in the horizontal rows are given by

$$h_I^* \colon \mathrm{Hom}_{\mathbb{C}}(\widetilde{V}_I, \mathbb{C}) \to \mathrm{Hom}_{\mathbb{C}}(V_I, \mathbb{C})$$
$$\tilde{\alpha}_I \mapsto \tilde{\alpha}_I \circ h_I$$

and

$$\mathrm{Hom}_{\mathbb{C}}((h_J), \mathcal{O}_{I,0}) \colon \bigoplus_J \mathrm{Hom}_{\mathbb{C}}(\widetilde{V}_J, \mathcal{O}_{I,0}) \to \bigoplus_J \mathrm{Hom}_{\mathbb{C}}(V_J, \mathcal{O}_{I,0})$$
$$\oplus_J \tilde{\alpha}_I^J \mapsto \oplus_J (\tilde{\alpha}_I^J \circ h_J)\,.$$

The isomorphisms from the lower row into the upper row are given by (ALG) composed with (η_I) and $(\tilde{\eta}_I)$, respectively (see part (ii) and part (v) of the proof of Proposition 4.6).
The commutativity of the diagram follows now easily using the commutativity conditions of the morphism (h_I) with the $B_{\bullet,\bullet}$ and $\tilde{B}_{\bullet,\bullet}$-maps (see Definition 1.1).

The functors D, Q and $\mathscr{A}$ are equivalences of categories (cf. Corollary 1.10, Theorem 1.11, Theorem 4.2 and the subsequent remark there). Hence, E is an equivalence of categories as well and, therefore, essentially surjective. Moreover, we know that D is its own quasi-inverse and $\mathcal{G}$ is the inverse of Q. This shows now that $D \circ \mathcal{G} \circ \mathscr{A}$ is a quasi-inverse of E. In the same manner we obtain that $E \circ D \circ \mathcal{G}$ is a quasi-inverse of $\mathscr{A}$. □

Remark 4.9.
In [KV06, Proposition 4.4] an equivalence of categories in the case of a central arrangement $\mathcal{C}$ of hyperplanes in $\mathbb{C}^n$ is also stated: Namely, one obtains an equivalence between a full subcategory of the category $\mathcal{Q}ui_{\mathcal{C}}$ of quiver representations corresponding to this arrangement (similarly defined in general as in the normal crossing case in Section 2.2) and a full subcategory closed under extensions of the category $\mathcal{M}od_h(\mathscr{D})$ of holonomic algebraic $\mathscr{D}$-modules. The full subcategory of the category of quiver representations $\mathcal{Q}ui_{\mathcal{C}}$ is defined by restricting the eigenvalues of several maps involved in the quiver representation. The full subcategory of $\mathcal{M}od_h(\mathscr{D})$ is not stated explicitly. The proof of Khoroshkin and Varchenko uses a gluing theorem of Beilinson-Malgrange-Kashiwara to glue quiver representations of "lower level" which allows them to use induction on n. On the other side this construction makes it somehow impossible to state the essential image of their equivalence explicitly. In contrast to that we give a complete classification of quiver $\mathscr{D}$-modules in the case of a normal crossing.

In dimension 1 their category of quiver representations with restricted eigenvalues is more or less the category $\mathcal{Q}ui_1^{\Sigma_1}$. Taking a quiver representation

$$V_\varnothing \underset{A_{\varnothing,\alpha}}{\overset{A_{\alpha,\varnothing}}{\rightleftarrows}} V_\alpha ,$$

they also use that the eigenvalues of $A_{\varnothing,\alpha}A_{\alpha,\varnothing}$ must be contained in a set of $\mathbb{C}^1$ where no two points differ by $\mathbb{Z}$. Accordingly the category $\mathcal{Q}ui_1^{\Sigma_1}$ would be admissible for their equivalence.

But already in dimension 2 one can easily see that their restrictions of eigenvalues are much more rigid than belonging to $\mathcal{Q}ui_2^{\Sigma_1}$. Given a quiver representation

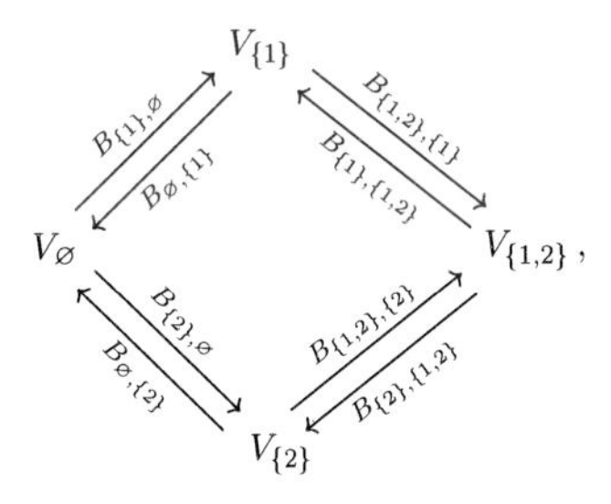

it would be admissible for them if for example the eigenvalues of

$$B_{\varnothing,\{1\}}B_{\{1\},\varnothing},\quad B_{\varnothing,\{2\}}B_{\{2\},\varnothing},\quad B_{\{1\},\{1,2\}}B_{\{1,2\},\{1\}},\quad B_{\{2\},\{1,2\}}B_{\{1,2\},\{2\}},$$
$$B_{\{1,2\},\{1\}}B_{\{1\},\{1,2\}}+B_{\{1,2\},\{2\}}B_{\{2\},\{1,2\}} \text{ and } B_{\varnothing,\{1\}}B_{\{1\},\varnothing}+B_{\varnothing,\{2\}}B_{\{2\},\varnothing}$$

are contained in Σ_1. The maps in the first line of this list are those we are using for $\mathcal{Q}ui_2^{\Sigma_1}$. But the ones in the second line do not need to fulfil additional properties in our case. This is a strong evidence that the essential image of their equivalence in our setting is not $\mathcal{M}od_{\mathrm{h}}^S(\mathscr{D})$ or $\mathcal{M}od_{\mathrm{rh}}^S(\mathscr{D})$ – where however they are using algebraic $\mathscr{D}$-modules.

Finally, we consider algebraic $\mathscr{D}$-modules or rather a connection between analytic and algebraic ones in our normal crossing case. In the following corollary we will see that every analytic $\mathscr{D}$-module from $\mathcal{M}od^S_{\mathrm{rh}}(\mathscr{D})$ – which we will denote $\mathcal{M}od^S_{\mathrm{rh}}(\mathscr{D}^{\mathrm{an}})$ in the following – is isomorphic to the analytization of an algebraic $\mathscr{D}$-module.

Corollary 4.10.
Let $\mathscr{M}^{an}$ be an object in $\mathcal{M}od^S_{rh}(\mathscr{D}^{an})$. Then, there exists an object $\mathscr{M}^{alg}$ in $\mathcal{M}od^S_{rh}(\mathscr{D}^{alg})$ such that $\mathscr{M}^{an}$ is isomorphic to the analytization of $\mathscr{M}^{alg}$, i. e.

$$\mathscr{M}^{an} \cong \mathcal{O}^{an}_{\mathbb{C}^n} \otimes_{\mathcal{O}^{alg}_{\mathbb{C}^n}} \mathscr{M}^{alg}.$$

The category $\mathcal{M}od^S_{\mathrm{rh}}(\mathscr{D}^{\mathrm{alg}})$ is the category of regular singular holonomic algebraic $\mathscr{D}^{\mathrm{alg}}$-modules whose characteristic variety is contained in Δ_S. The concepts of characteristic variety and holonomicity are defined analogously for algebraic $\mathscr{D}^{\mathrm{alg}}$-modules as they are for analytic $\mathscr{D}^{\mathrm{an}}$-modules (cf. for instance [BEG+87, Section VI.1]). We say that a holonomic algebraic $\mathscr{D}^{\mathrm{alg}}$-module $\mathscr{M}$ is regular iff $(j_*\mathscr{M})^{\mathrm{an}}$ is regular where $j\colon X \hookrightarrow \overline{X}$ is a compactification of X (see [Meb89, Subsection II.8.4] or [HTT08, Section 6.1] for this definition of regularity in the algebraic case).

Proof:
Clearly, in Definition 2.1 one might replace $\mathscr{D} = \mathscr{D}^{\mathrm{an}}$ by $\mathscr{D}^{\mathrm{alg}}$ which gives us the definition of an algebraic quiver $\mathscr{D}$-module as used by Khoroshkin and Varchenko. Adopting the proof of Corollary 2.1 and Proposition/Definition 2.2 we obtain a functor E^{alg}. Using the proof of Theorem 2.4 we directly see that this yields a functor E^{alg} from the category $\mathcal{Q}ui_n$ to the category of holonomic $\mathscr{D}^{\mathrm{alg}}$-modules whose characteristic variety is contained in Δ_S. The only point which needs a little more work is the regularity of a given algebraic quiver $\mathscr{D}$-module $E^{\mathrm{alg}}\mathcal{V}_n$. The regularity of $(j_*E^{\mathrm{alg}}\mathcal{V}_n)^{\mathrm{an}} = \mathcal{O}^{\mathrm{an}}_{\mathbb{C}^n} \otimes_{\mathcal{O}^{\mathrm{alg}}_{\mathbb{C}^n}} j_*E^{\mathrm{alg}}\mathcal{V}_n$ for any section which does not contain infinity is clear by just copying the proof of Theorem 2.4 as these sections of $j_*E^{\mathrm{alg}}\mathcal{V}_n$ are exactly the ones of $E^{\mathrm{alg}}\mathcal{V}_n$. Moreover, $(j_*E^{\mathrm{alg}}\mathcal{V}_n)^{\mathrm{an}}$ is regular at infinity if it was already regular near infinity using the definition of j_*. We choose coordinates at infinity by $t_i = \frac{1}{z_i}$. But as $t_i\partial_{t_i} = -z_i\partial_{z_i}$, we can copy the proof of Theorem 2.4 again. All in all, we obtain a functor E^{alg} from $\mathcal{Q}ui_n$ to $\mathcal{M}od^S_{\mathrm{rh}}(\mathscr{D}^{\mathrm{alg}})$ – which corresponds to the functor E in [KV06].

Given an analytic $\mathscr{D}$-module $\mathscr{M}^{\mathrm{an}}$, Theorem 4.5 shows us that $\mathscr{M}^{\mathrm{an}}$ is isomorphic to an analytic quiver $\mathscr{D}$-module $\mathscr{M}_Q = E^{\mathrm{an}}\mathcal{V}_n$ (as given in Definition 2.1) corresponding to the quiver representation $\mathcal{V}_n$ in $\mathcal{Q}ui_n^{\Sigma_1}$. Use the same quiver representation $\mathcal{V}_n$ in the algebraic context to define the algebraic quiver $\mathscr{D}$-module $\mathscr{M}_{Qalg} = E^{\mathrm{alg}}\mathcal{V}_n$. The equality $\mathscr{D}^{\mathrm{an}} = \mathcal{O}^{\mathrm{an}}_{\mathbb{C}^n} \otimes_{\mathcal{O}^{\mathrm{alg}}_{\mathbb{C}^n}} \mathscr{D}^{\mathrm{alg}}$ and right-exactness of tensoring yields now $\mathscr{M}_Q = \mathcal{O}^{\mathrm{an}}_{\mathbb{C}^n} \otimes_{\mathcal{O}^{\mathrm{alg}}_{\mathbb{C}^n}} \mathscr{M}_{Qalg}$. This proves the claim. $\square$

Bibliography

[AB94] D. V. Anosov and A. A. Bolibruch. *The Riemann-Hilbert problem.* Aspects of Mathematics, E 22. Friedr. Vieweg & Sohn, Braunschweig, 1994.

[BEG+87] A. Borel, F. Ehlers, P.-P. Grivel, A. Haefliger, B. Kaup, and B. Malgrange. *Algebraic D-modules*, volume 2 of *Perspectives in mathematics.* Academic Press, Inc., Boston, 1987.

[Bjö93] J.-E. Björk. *Analytic $\mathscr{D}$-modules and applications*, volume 247 of *Mathematics and its applications.* Kluwer Academic Publishers, Dordrecht, 1993.

[Bou83] L. Boutet de Monvel. $\mathscr{D}$-modules holonômes réguliers en une variable. In *Mathématique et physique. séminaire de l'Ecole Normale Supérieure, 1979–1982*, volume 37 of *Progress in mathematics*, pages 313–321. Birkhäuser, Boston, Mass., 1983.

[Del70] P. Deligne. *Equations différentielles à points singuliers réguliers*, volume 163 of *Lecture notes in mathematics.* Springer-Verlag, Berlin, 1970.

[Dim04] A. Dimca. *Sheaves in topology.* Universitext. Springer-Verlag, Berlin, 2004.

[Fis11] G. Fischer. *Lernbuch Lineare Algebra und Analytische Geometrie.* Studium. Vieweg+Teubner, Wiesbaden, 1st edition, 2011.

[GGM85a] A. Galligo, M. Granger, and Ph. Maisonobe. $\mathscr{D}$-modules et faisceaux pervers dont le support singulier est un croisement normal. *Ann. Inst. Fourier (Grenoble)*, 35(1):1–48, 1985.

[GGM85b] A. Galligo, M. Granger, and Ph. Maisonobe. $\mathscr{D}$-modules et faisceaux pervers dont le support singulier est un croisement normal. II. *Astérisque*, (130):240–259, 1985.

[GM93] M. Granger and Ph. Maisonobe. A basic course on differential modules. In *Éléments de la théorie des systèmes différentiels. $\mathscr{D}$-modules cohérents et holonomes*, volume 45 of *Travaux en Cours*, pages 103–168. Hermann, Paris, 1993.

[Gra00] J. J. Gray. *The Hilbert challenge.* Oxford University Press, Oxford, 2000.

[Hil02] D. Hilbert. Mathematical problems. Lecture delivered before the international congress of mathematicians at Paris in 1900. *Bull. Amer. Math. Soc.*, 8(10):437–479, 1902. Originally published as *Mathematische Probleme. Vortrag, gehalten auf dem internationalen Mathematiker-Kongreß. Nachrichten der Akademie der Wissenschaften in Göttingen, 1900.* Translated for the Bulletin, with the author's permission, by Dr. Mary Winston Newson.

[HJ91] R. A. Horn and C. R. Johnson. *Topics in matrix analysis.* Cambridge University Press, Cambridge, 1991.

[HTT08] R. Hotta, K. Takeuchi, and T. Tanisaki. *D-modules, perverse sheaves, and representation theory*, volume 236 of *Progress in mathematics*. Birkhäuser, Boston, Mass., 2008.

[Kas84] M. Kashiwara. The Riemann-Hilbert problem for holonomic systems. *Publ. Res. Inst. Math. Sci.*, 20(2):319–365, 1984.

[Kas03] M. Kashiwara. *D-modules and microlocal calculus*, volume 217 of *Translations of mathematical monographs*. American Mathematical Society, Providence, R.I., 2003.

[Kho95] S. Khoroshkin. $\mathscr{D}$-modules over the arrangements of hyperplanes. *Comm. Algebra*, 23(9):3481–3504, 1995.

[KS98] S. Khoroshkin and V. Schechtman. Non-resonance D-modules over arrangements of hyperplanes. `http://arxiv.org/abs/math/9801134v3`, 1998.

[KV06] S. Khoroshkin and A. Varchenko. Quiver D-modules and homology of local systems over an arrangement of hyperplanes. *IMRP Int. Math. Res. Pap.*, 2006:1–116, 2006. Art. ID 69590.

[Mal91] B. Malgrange. *Equations différentielles à coefficients polynomiaux*, volume 96 of *Progress in mathematics*. Birkhäuser, Boston, Mass., 1991.

[Meb84] Z. Mebkhout. Une équivalence de catégories, Une autre équivalence de catégories. *Compos. Math.*, 51(1):51–62, 63–88, 1984.

[Meb89] Z. Mebkhout. *Le formalisme des six opérations de Grothendieck pour les* $\mathscr{D}_X$*-modules cohérents*, volume 35 of *Travaux en Cours*. Hermann, Paris, 1989.

[MM04] Ph. Maisonobe and Z. Mebkhout. Le théorème de comparaison pour les cycles évanescents. In *Éléments de la théorie des systèmes différentiels géométriques*, volume 8 of *Séminaires et Congrès*, pages 311–389. Société Mathématique de France, 2004.

[Ple08] J. Plemelj. Riemannsche Funktionenscharen mit gegebener Monodromiegruppe. *Monatsh. Math. Phys.*, 19(1):211–245, 1908.

[Sab93] C. Sabbah. Introduction to algebraic theory of linear systems of differential equations. In *Éléments de la théorie des systèmes différentiels.* $\mathscr{D}$*-modules cohérents et holonomes*, volume 45 of *Travaux en Cours*, pages 1–80. Hermann, Paris, 1993.

[Ser75] J.-P. Serre. *Algèbre locale, multiplicités*, volume 11 of *Lecture notes in mathematics*. Springer-Verlag, Berlin, 3rd edition, 1975.

[Thi05] R. Thiele. Hilbert and his twenty-four problems. In *Mathematics and the historian's craft*, volume 21 of *CMS books in mathematics*, pages 243–295. Springer, New York, 2005.

[Yan02] B. H. Yandell. *The honors class. Hilbert's problems and their solvers.* A K Peters, Natick, Mass., 2002.

Lebenslauf

Persönliche Angaben

Name	Stephanie Zapf, geb. Winhart
Geburtsdaten	9. Mai 1987 in Kösching, Kreis Eichstätt

Ausbildung

1997 – 2006	Descartes-Gymnasium, Neuburg a. d. Donau
2006 – 2011	Studium der Mathematik, Abschluss Diplom-Mathematikerin, Universität Augsburg
2011 – 2015	Promotion am Lehrstuhl für Algebra und Zahlentheorie, Universität Augsburg

Kontakt

Universität Augsburg
Lehrstuhl für Algebra und Zahlentheorie
D-86153 Augsburg

Augsburg, August 2015

In der Reihe *Augsburger Schriften zur Mathematik, Physik und Informatik,*
herausgegeben von Prof. Dr. B. Aulbach, Prof. Dr. F. Pukelsheim,
Prof. Dr. W. Reif, Prof. Dr. B. Schmidt, Prof. Dr. D. Vollhardt,
sind bisher erschienen:

1	Martin Mißlbeck	Entwicklung eines schnellen Spektralradiometers und Weiterentwicklung herkömmlicher Messverfahren zur Messung der solaren UV-Strahlung	
		ISBN 978-3-8325-0208-9, 2003, 139 S.	40.50 €
2	Bernd Reinhard	Dynamisches Trapping in modulierten monotonen Potentialen	
		ISBN 978-3-8325-0516-5, 2004, 154 S.	40.50 €
3	Cosima Schuster	Physikerinnen stellen sich vor - Dokumentation der Deutschen Physikerinnentagung 2003	
		ISBN 978-3-8325-0520-2, 2004, 164 S.	40.50 €
4	Udo Schwingenschlögl	The Interplay of Structural and Electronic Properties in Transition Metal Oxides	
		ISBN 978-3-8325-0530-1, 2004, 174 S.	40.50 €
5	Marianne Leitner	Zero Field Hall-Effekt für Teilchen mit Spin 1/2	
		ISBN 978-3-8325-0578-3, 2004, 81 S.	40.50 €
6	Georg Keller	Realistic Modeling of Strongly Correlated Electron Systems	
		ISBN 978-3-8325-0970-5, 2005, 150 S.	40.50 €
7	Niko Tzoukmanis	Local Minimizers of Singularly Perturbed Functionals with Nonlocal Term	
		ISBN 978-3-8325-0650-6, 2004, 154 S.	40.50 €

8 Stefan Bettner

Beweis der Kongruenzen von Berwick sowie deren Verallgemeinerung und weitere Anwendungen von Torsionspunkten auf elliptischen Kurven

ISBN 978-3-8325-0709-1, 2004, 150 S. 40.50 €

9 Ulrich Miller

Rigorous Numerics using Conley Index Theory

ISBN 978-3-8325-0854-8, 2005, 130 S. 40.50 €

10 Markus Lilli

The Effect of a Singular Perturbation to a 1-d Non-Convex Variational Problem

ISBN 978-3-8325-0928-6, 2005, 100 S. 40.50 €

11 Stefan Krömer

The Effect of a Singular Perturbation to a 1-d Non-Convex Variational Problem

ISBN 978-3-8325-0928-6, 2005, 100 S. 40.50 €

12 Andreas Nickel

The Effect of a Singular Perturbation to a 1-d Non-Convex The Lifted Root Number Conjecture for small sets of places and an application to CM-extensions

ISBN 978-3-8325-1969-8, 2008, 102 S. 33.00 €

13 Torben Stender

Growth Rates for Semiflows with Application to Rotation Numbers for Control Systems

ISBN 978-3-8325-2165-3, 2009, 136 S. 34.50 €

14 Tobias Wichtrey

Harmonic Limits of Dynamical and Control Systems

ISBN 978-3-8325-2796-9, 2011, 207 S. 37.00 €

15 Christian A. Möller

Adaptive Finite Elements in the Discretization of Parabolic Problems

ISBN 978-3-8325-2815-7, 2011, 257 S. 37.00 €

16 Georg M. Reuther

Decoherence and Time-Resolved Readout in Superconducting Quantum Circuits

ISBN 978-3-8325-2846-1, 2011, 157 S. 35.00 €

17 Ralf Seger

Exploratory Model Comparison

ISBN 978-3-8325-2927-7, 2011, 205 S. 60.00 €

18	Olga Birkmeier	Machtindizes und Fairness-Kriterien in gewichteten Abstimmungssystemen mit Enthaltungen ISBN 978-3-8325-2968-0, 2011, 153 S. 37.50 €
19	Johannes Neher	A posteriori error estimation for hybridized mixed and discontinuous Galerkin methods ISBN 978-3-8325-3088-4, 2012, 105 S. 33.50 €
20	Andreas Krug	Extension groups of tautological sheaves on Hilbert schemes of points on surfaces ISBN 978-3-8325-3254-3, 2012, 127 S. 34.00 €
21	Isabella Graf	Multiscale modeling and homogenization of reaction-diffusion systems involving biological surfaces ISBN 978-3-8325-3397-7, 2013, 285 S. 46.50 €
22	Franz Vogler	Derived Manifolds from Functors of Points ISBN 978-3-8325-3405-9, 2013, 158 S. 35.00 €
23	Kai-Friederike Oelbermann	Biproportionale Divisormethoden und der Algorithmus der alternierenden Skalierung ISBN 978-3-8325-3456-1, 2013, 91 S. 32.50 €
24	Markus Göhl	Der durchschnittliche Rechenaufwand des Simplexverfahrens unter einem verallgemeinerten Rotationssymmetriemodell ISBN 978-3-8325-3531-5, 2013, 141 S. 34.50 €
25	Fabian Reffel	Konvergenzverhalten des iterativen proportionalen Anpassungsverfahrens im Fall kontinuierlicher Maße und im Fall diskreter Maße ISBN 978-3-8325-3652-7, 2014, 185 S. 40.00 €
26	Emanuel Schnalzger	Lineare Optimierung mit dem Schatteneckenalgorithmus im Kontext probabilistischer Analysen ISBN 978-3-8325-3788-3, 2014, 225 S. 42.00 €
27	Robert Gelb	D-modules: Local formal convolution of elementary formal meromorphic connections ISBN 978-3-8325-3894-1, 2015, 95 S. 33.00 €

28	Manuel Friedrich	Effective Theories for Brittle Materials: A Derivation of Cleavage Laws and Linearized Griffith Energies from Atomistic and Continuum Nonlinear Models ISBN 978-3-8325-4028-9, 2015, 291 S. 39.50 €
29	Hedwig Heizinger	Stokes Structure and Direct Image of Irregular Singular $\mathcal{D}$-Modules ISBN 978-3-8325-4061-6, 2015, 70 S. 37.00 €
30	Stephanie Zapf	Quiver $\mathcal{D}$-Modules and the Riemann-Hilbert Correspondence ISBN 978-3-8325-4084-5, 2015, 80 S. 32.50 €

Alle erschienenen Bücher können unter der angegebenen ISBN im Buchhandel oder direkt beim Logos Verlag Berlin (www.logos-verlag.de, Fax: 030 - 42 85 10 92) bestellt werden.